Contents

KU-106-959

Introduction to the Teacher

A First Course in Technical English is a highly practical, elementary course for foreign students needing English for technical studies. The starting point assumes a low level of language acquisition and is suitable for near-Beginners. Book 1 is divided into eight units which deal with different technical topics and each unit is sub-divided into three sections. Each section is based on a series of illustrated texts which highlight a limited number of language features and these are practised through the exercises. Language items are selected for their relevance to the students' needs in using technical language. The content of the course is not too 'technical' since the aim is to teach English in a technical context, not technology through English.

Teacher's Book
A separate Teacher's Book is available to accompany Students' Book 1 and contains a general Introduction with detailed lesson-by-lesson notes. For the course to be used most effectively, the teacher will need to refer to the guidelines contained in the Teacher's Book.

Recorded Material
All the texts which begin each section have been recorded and are available on tape, with directions for use in the Teacher's Book.

£1-75
385512A
H2

A First Course in Technical English

Students' Book 1

LYNETTE BEARDWOOD
HUGH TEMPLETON
MARTIN WEBBER

Illustrations by
George Hartfield Ltd

HEINEMANN EDUCATIONAL BOOKS
LONDON

Heinemann Educational Books Ltd
LONDON EDINBURGH MELBOURNE AUCKLAND TORONTO
SINGAPORE HONG KONG KUALA LUMPUR
IBADAN NAIROBI JOHANNESBURG
NEW DELHI LUSAKA KINGSTON

ISBN 0 435 28755 9

[Teacher's Book ISBN 0 435 28756 7]
[Tape ISBN 0 435 28757 5]
[Cassette ISBN 0 435 28030 9]

© Lynette Beardwood, Hugh Templeton,
Martin Webber 1978

First published 1978

Cover photograph
by kind permission of
The British Steel Corporation

Published by
Heinemann Educational Books Ltd,
48 Charles Street, London W1X 8AH
Printed and bound in Great Britain by
Butler and Tanner Ltd., Frome and London

UNIT ONE Tools, Materials, Containers and Instruments

SECTION A: TOOLS

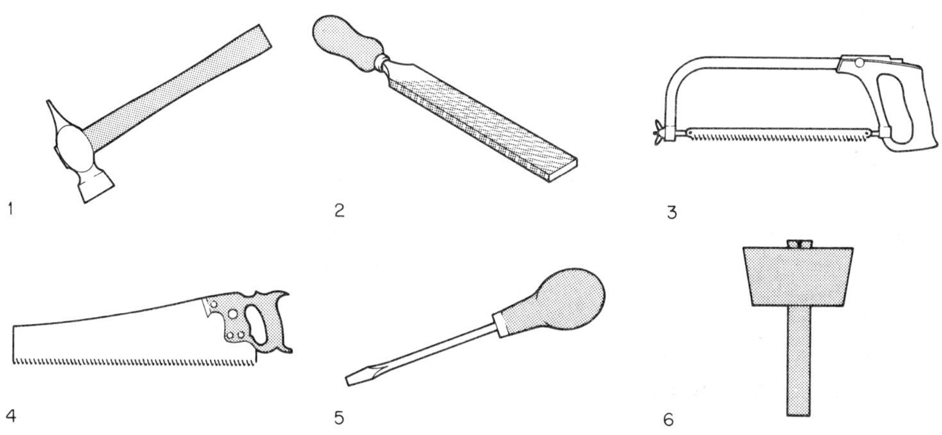

The first tool is a hammer. It has a handle and a head.

The second tool is a file. It has a handle and a blade.

The third tool is a hacksaw. It has a handle and a blade.

Exercise One Look at pictures four, five and six. Complete the sentences.

The fourth tool is a handsaw.
It has a . . . . and a . . . .

The fifth . . . . screwdriver.
It has . . . .

The sixth . . . . mallet.
. . . .

Exercise Two Complete this table.

	NAME OF PART		
NAME OF TOOL	a handle	a head	a blade
a hammer	√	√	X
a screwdriver			
a handsaw			
a file			
a mallet			
a hacksaw			

Look at this example:

A hammer has a handle and a head.

Make five sentences from the table in the same way.

Exercise Three

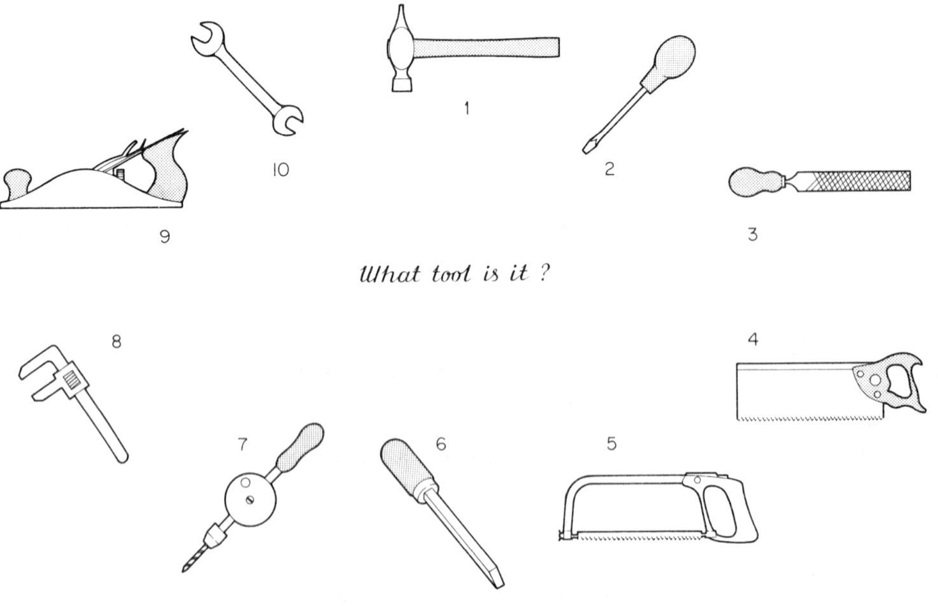

Exercise Four Look at the pictures and answer the questions.

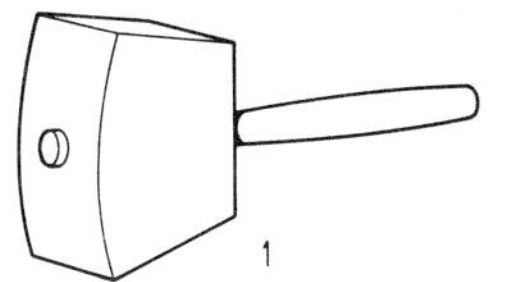

1. Is it a hammer or a mallet?

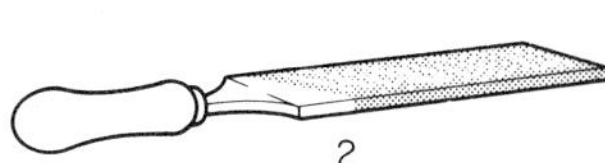

2. Is it a file or a chisel?

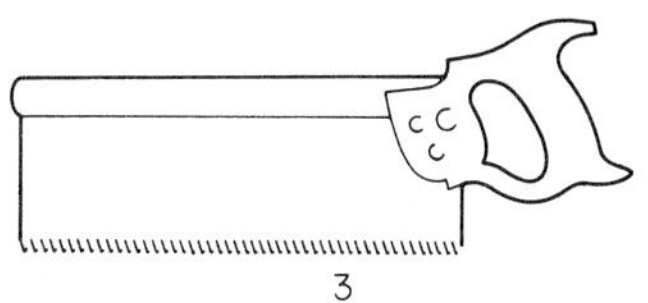

3. Is it a handsaw or a hacksaw?

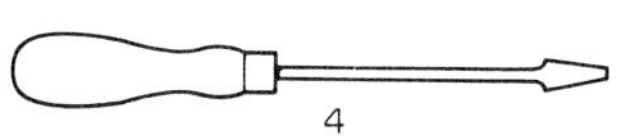

4. Is it a screwdriver or a drill?

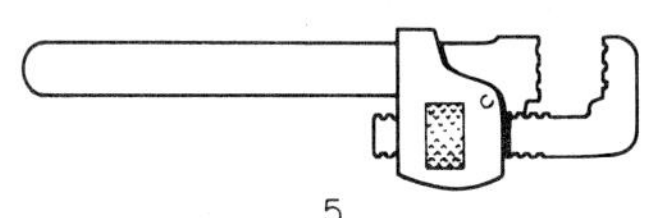

5. Is it a spanner or a wrench?

Exercise Five Complete the table.

1	one	first	1st
2			
3			
4			
5			
6			
7			
8			
9			
10			

a tool
a hammer
a file
a hacksaw
a handsaw

a screwdriver
a mallet
a chisel
a drill
a wrench

a plane
a spanner
a handle
a head
a blade

a name

a part

SECTION B: MATERIALS AND CONTAINERS

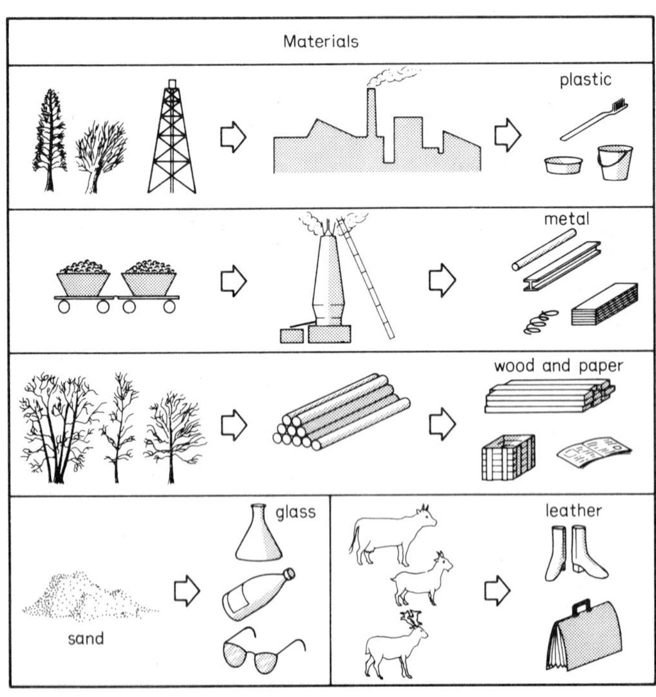

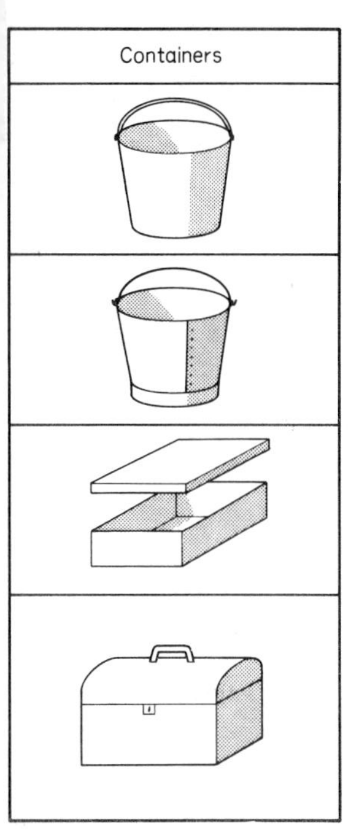

Plastic is a material. Metal is also a material.

The first container is a bucket. It is made of plastic. The bucket has a handle. The handle is also made of plastic.

The second container is another bucket. It is a metal bucket. It has a metal handle.

The third container is a paper box. The lid is also made of paper.

The fourth container is another box. It is made of wood. This box has a wooden handle and a metal lock.

Exercise Six

Look at the pictures and complete the sentences.

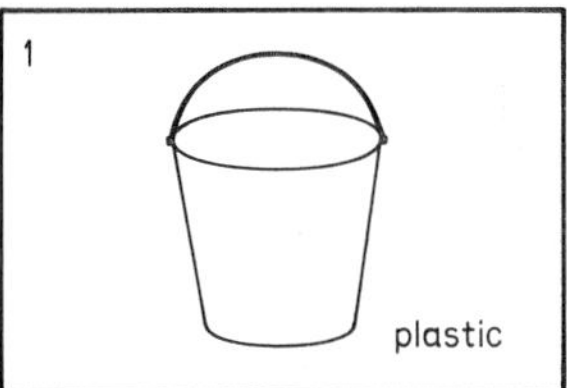

This container is a . . . .
It is a . . . . bucket.
The . . . . is also made of . . . .

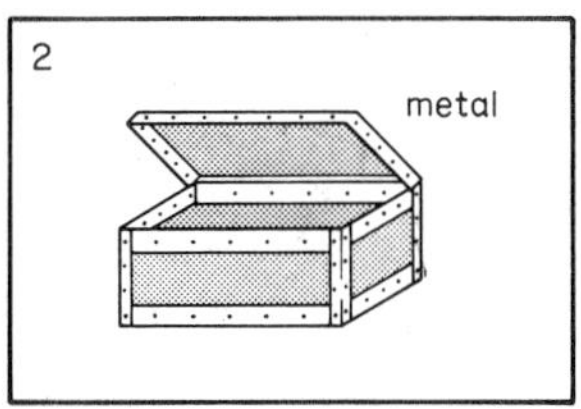

This . . . . is a . . . .
It is made . . . .
The . . . . is also . . . .

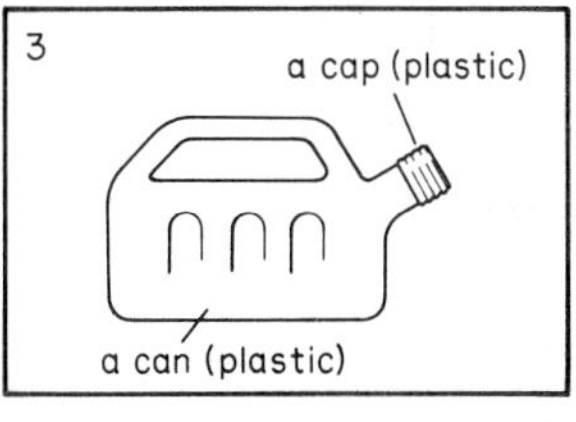

This . . . . can.
It . . . . can.
The cap . . . .

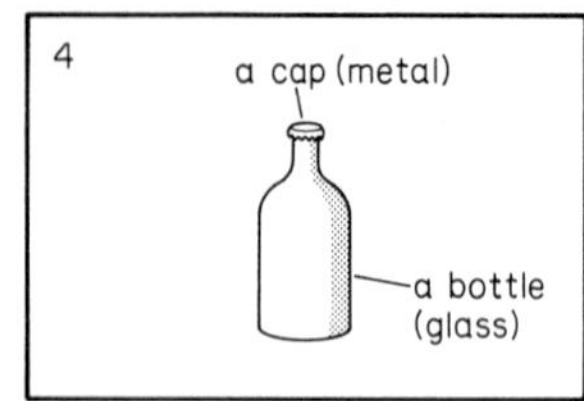

. . . . bottle.
. . . . glass.
It has . . . . cap.

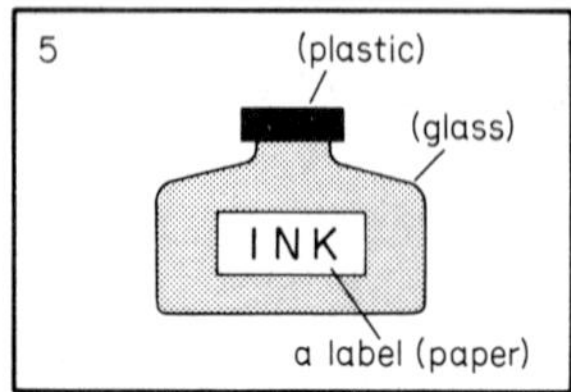

. . . . another . . . .
. . . . glass . . . .
The cap . . . . made . . . . and the
label . . . . made . . . .

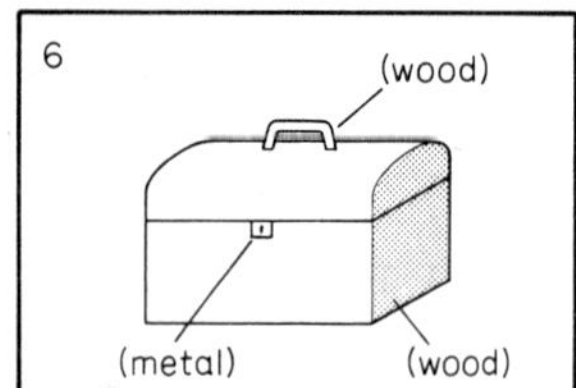

. . . . another . . . .
. . . . wooden . . . .
It has . . . . and . . . .
The . . . . made . . . . and . . . .
made . . . .

Exercise Seven

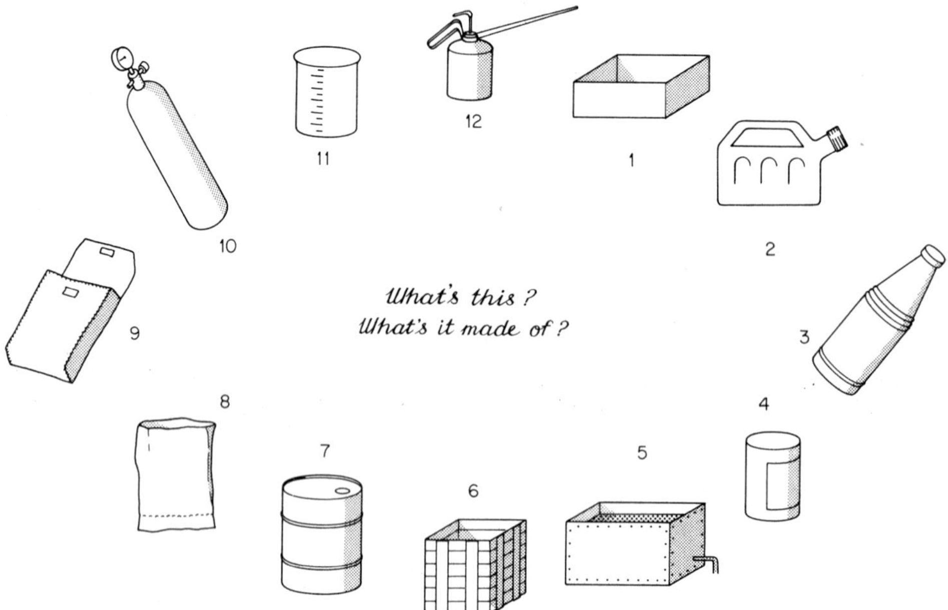

Exercise Eight

Look at the example.

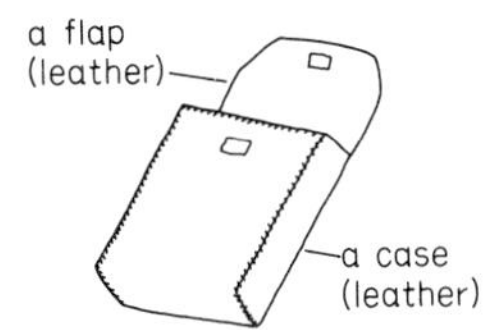

This case is made of leather.
It has a leather flap.

Make sentences from the pictures in the same way.

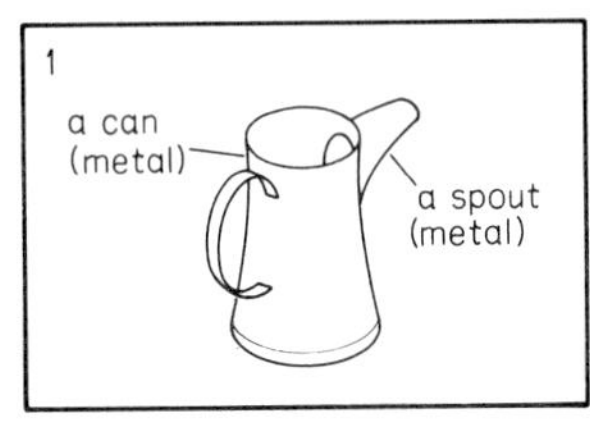

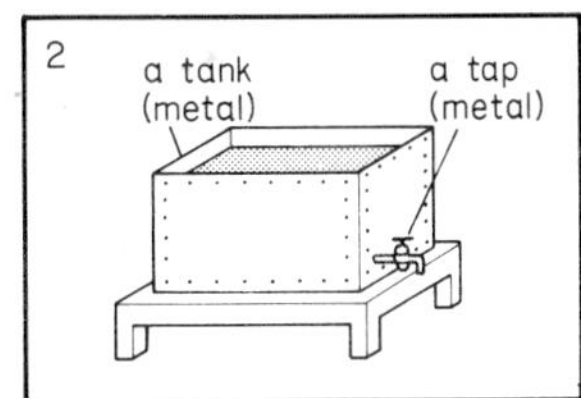

Now look at this example.

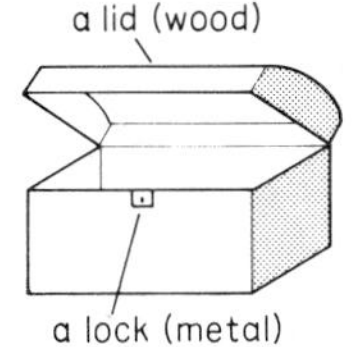

This box has a wooden lid and a metal lock. The lid is made of wood and the lock is made of metal.

Label the pictures and make sentences in the same way.

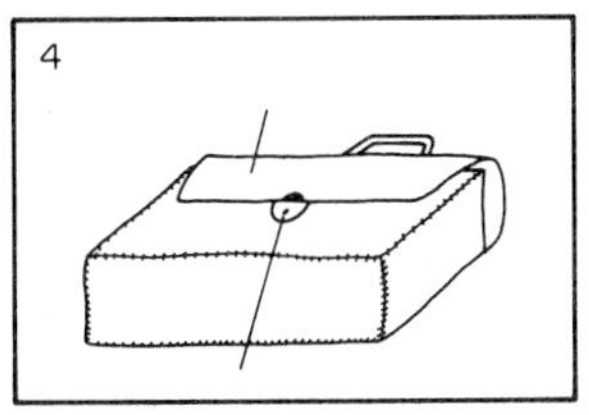

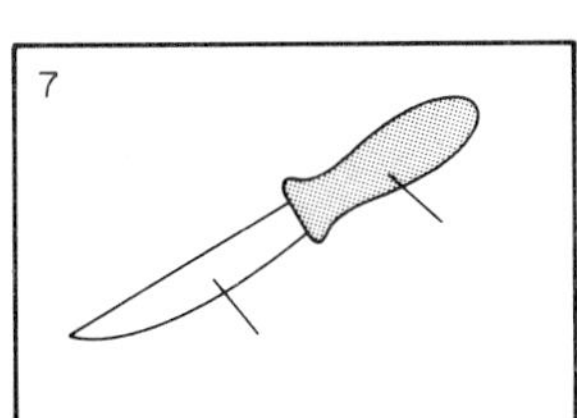

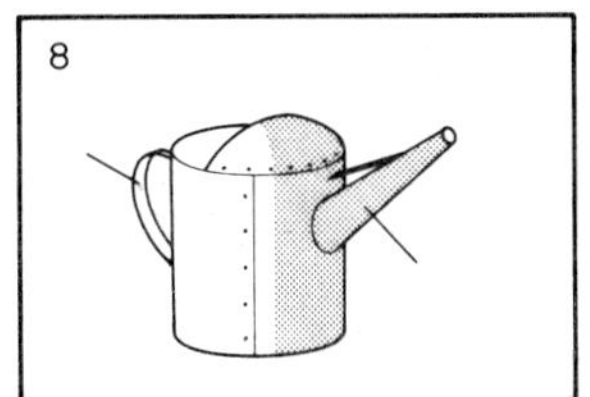

Exercise Nine

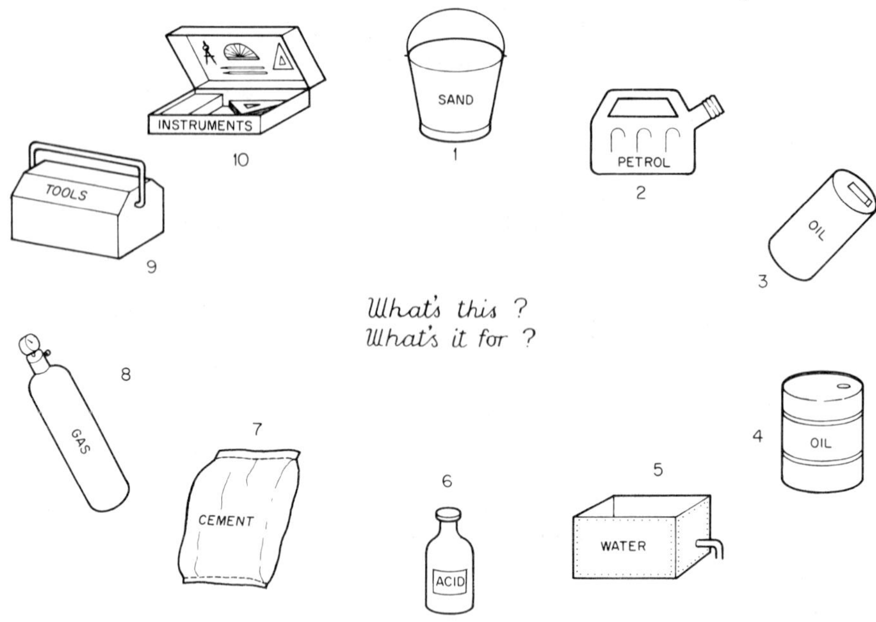

Exercise Ten Complete this table.

11	eleven	eleventh	11th
12			
13			
14			
15			
16			
17			
18			
19			
20			

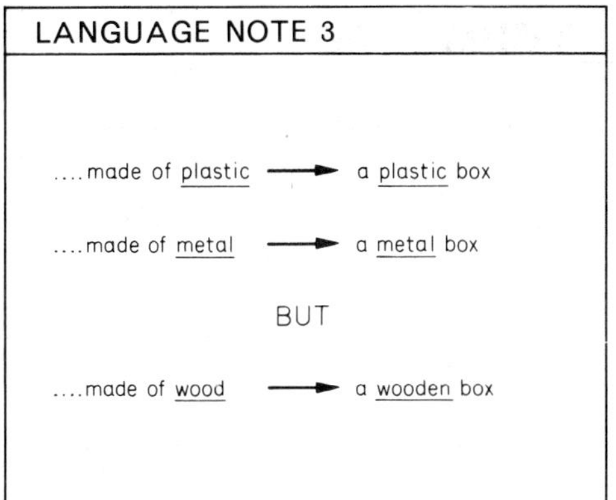

a material
plastic
metal
paper
wood (wooden)
glass
leather
sand
cement
petrol
oil
acid
water
gas

a container
a bucket
a box
a can
a bottle
a tank
a case
a drum
a sack
a cylinder
a beaker

a lid
a lock
a cap
a label
a flap
a spout
a tap

a knife

also

another

made of

SECTION C: INSTRUMENTS

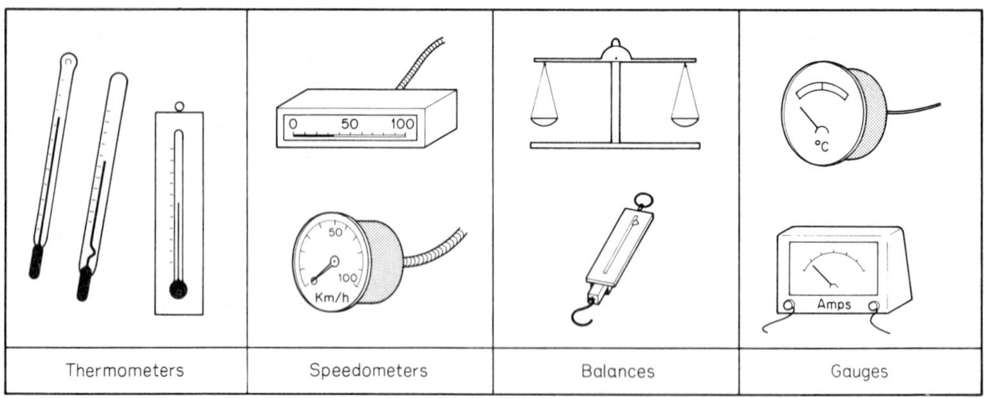

A thermometer is an instrument. It is used for measuring temperatures.

A speedometer is another instrument. It is used for measuring speeds.

Balances are instruments. They are used for measuring weights.

Gauges are also instruments. They are used for measuring temperatures or pressures.

Exercise Eleven Look at the example.

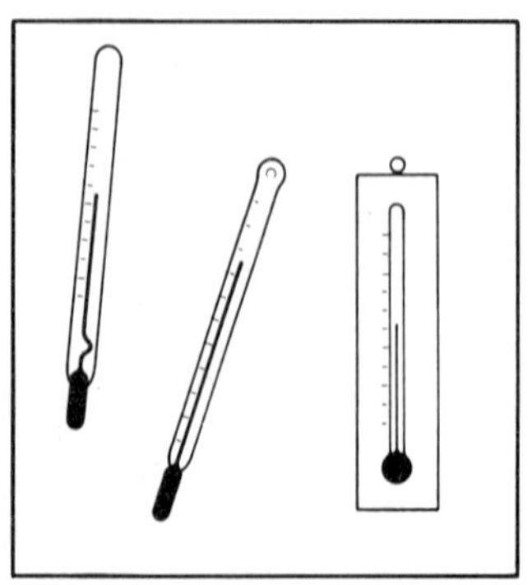

A thermometer is used for measuring temperatures.
Thermometers are used for measuring temperatures.

These two sentences have the same meaning.

Look at the pictures and complete the sentences.

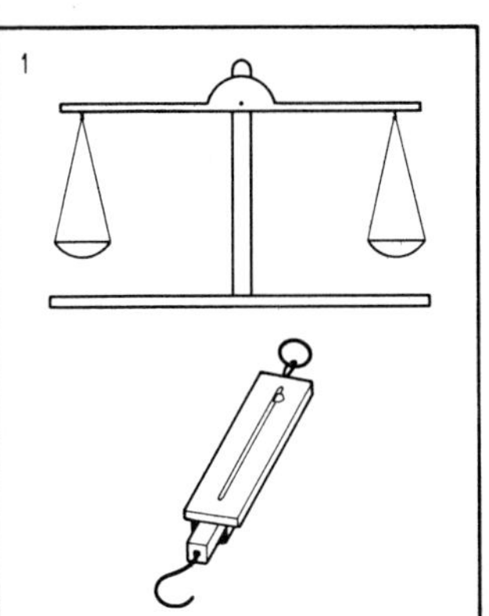

A . . . . is an . . . .
It . . . . measuring
. . . .

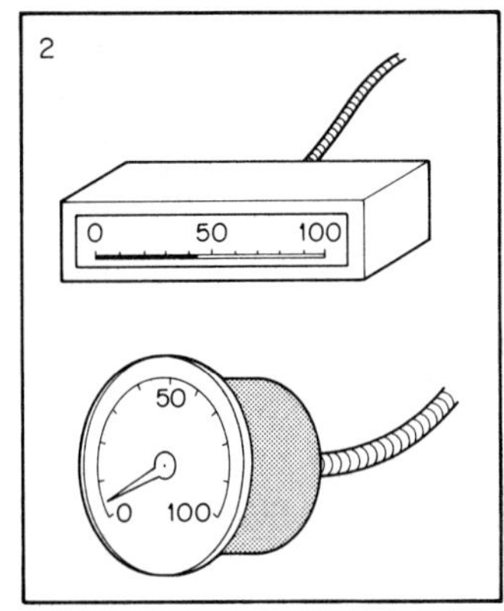

. . . . are . . . .
They . . . .
measuring . . . .

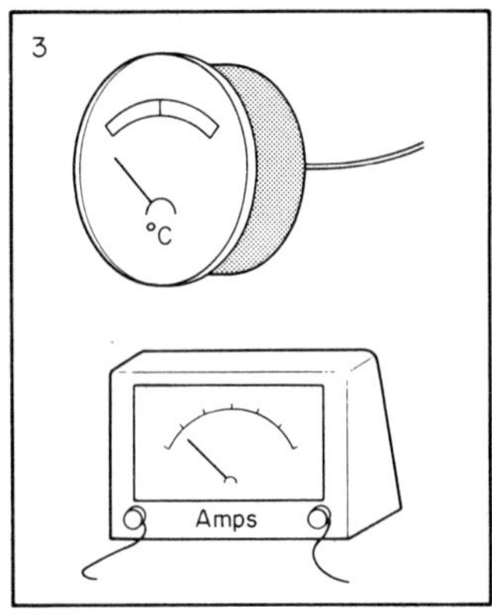

. . . .
. . . . temperatures
or pressures.

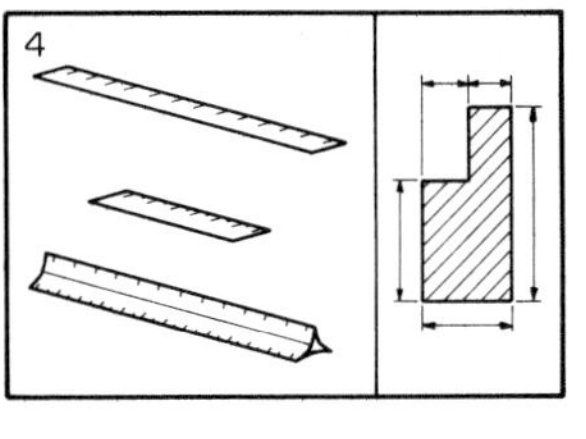

. . . .
. . . . dimensions.

. . . .
. . . . angles.

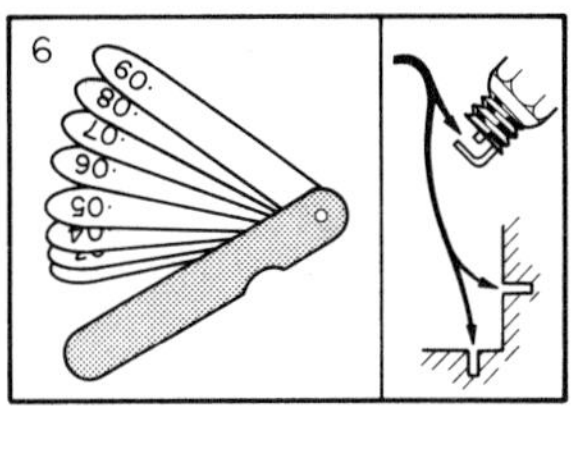

. . . .
. . . . gaps.

Exercise Twelve Look at the example.

This drum is used for holding petrol.

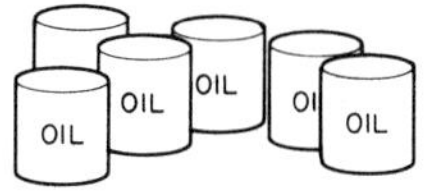

These drums are used for holding oil.

Look at the pictures and make sentences in the same way.

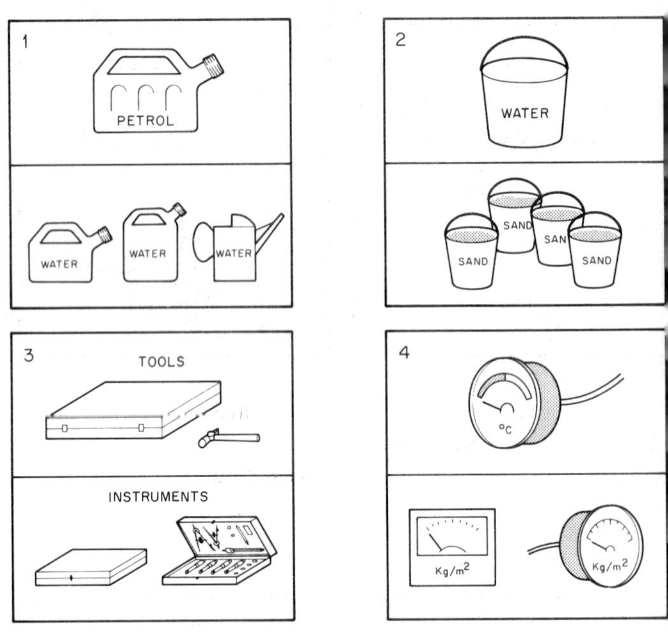

Exercise Thirteen

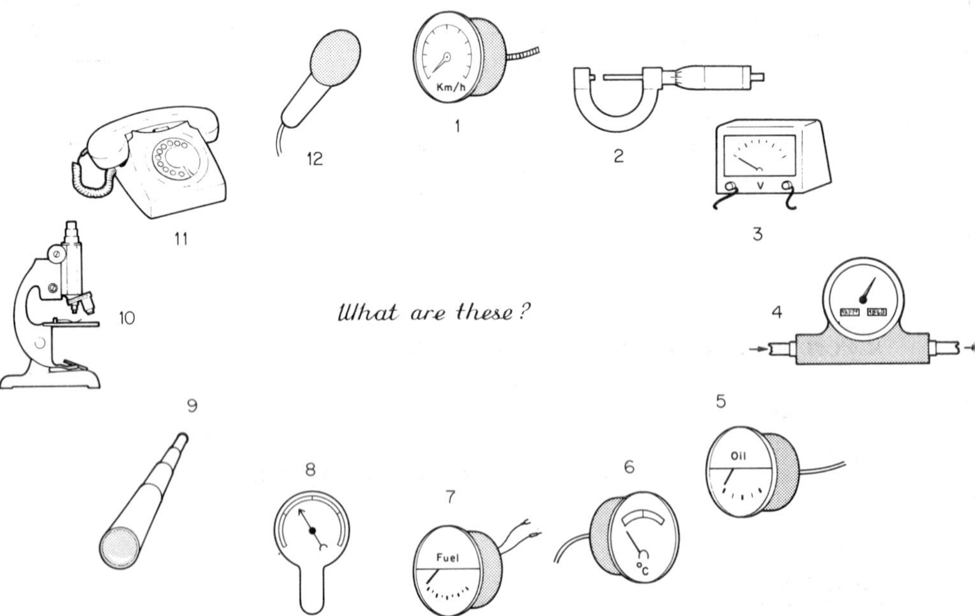

Exercise Fourteen What are these? What are they used for?

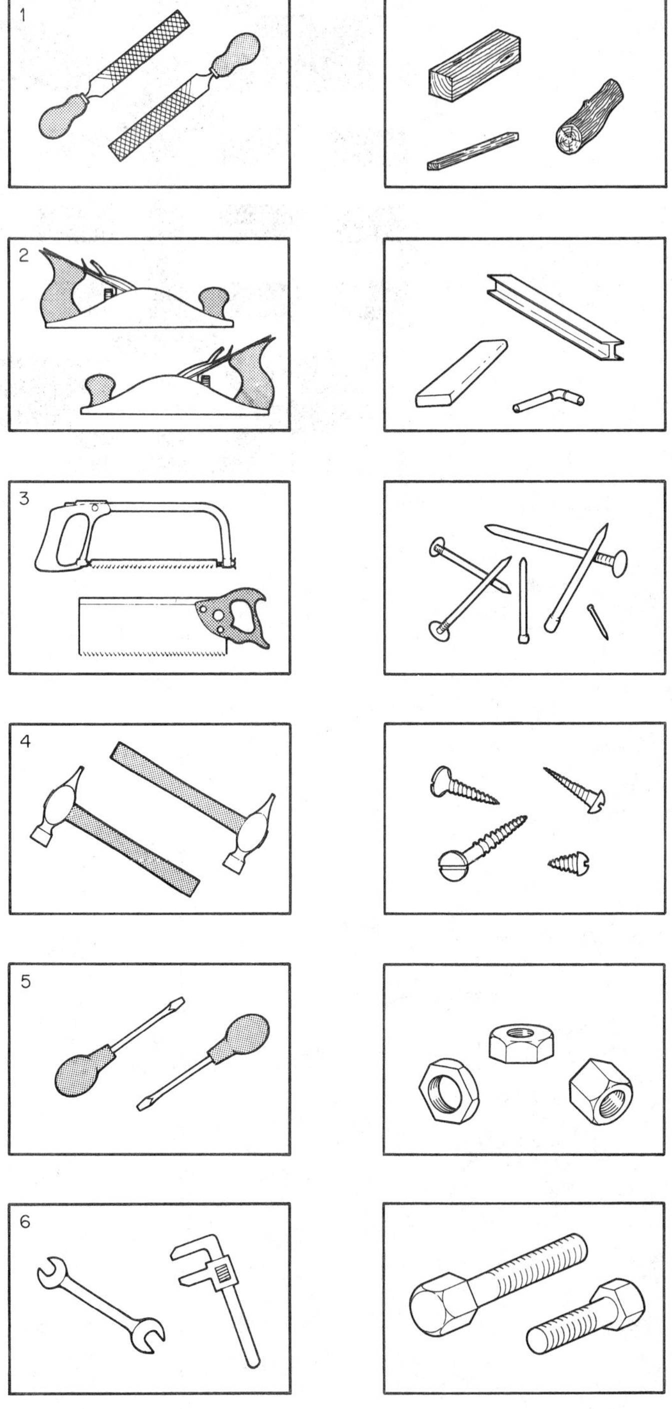

Exercise Fifteen Complete this crossword.

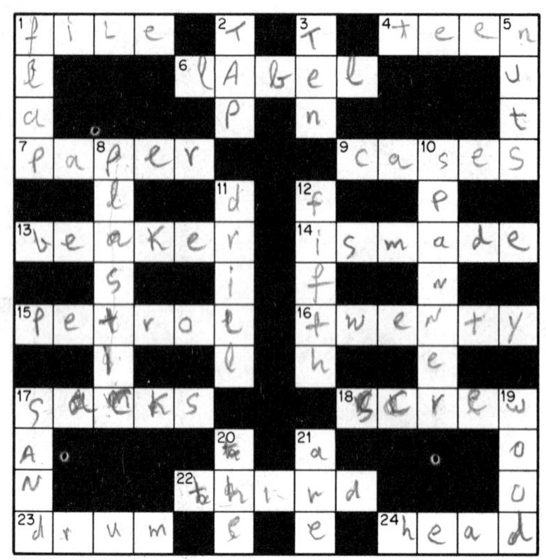

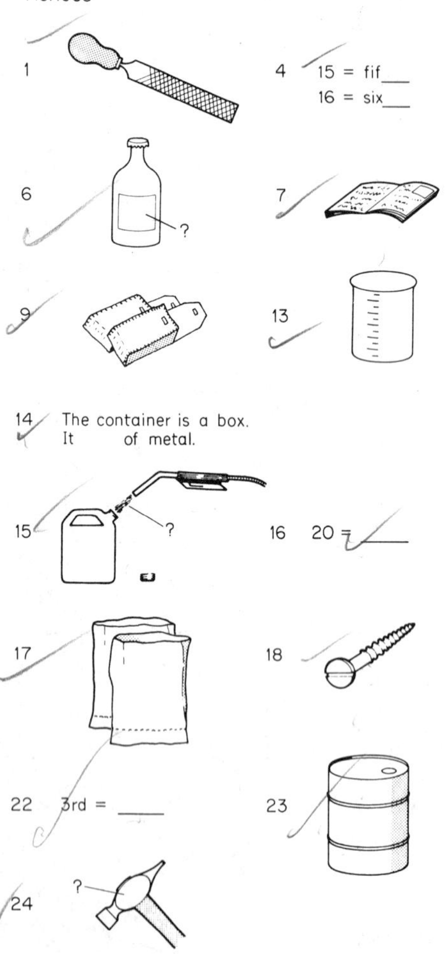

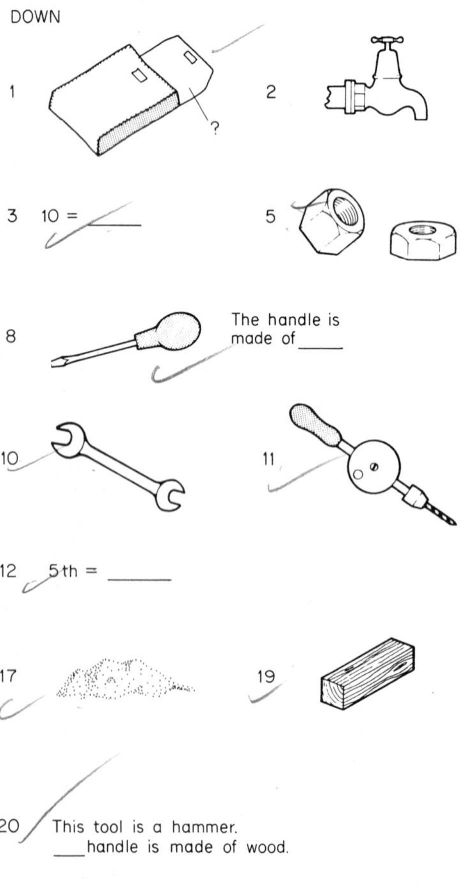

LANGUAGE NOTE 5		
a tool	BUT	an instrument
a dimension	BUT	an angle
a petrol gauge	BUT	an oil gauge

LANGUAGE NOTE 6

an instrument	a temperature	used for
a thermometer	a speed	measure
a speedometer	a weight	hold
a balance	a pressure	file
a gauge	a dimension	plane
a ruler	an angle	saw
a protractor	a gap	hammer
a feeler gauge	a nail	turn
a micrometer	a screw	
a voltmeter	a nut	
a rain gauge	a bolt	
a telescope		
a microscope		
a telephone		
a microphone		

UNIT TWO Geometrical Shapes

SECTION A: PLANE SHAPES

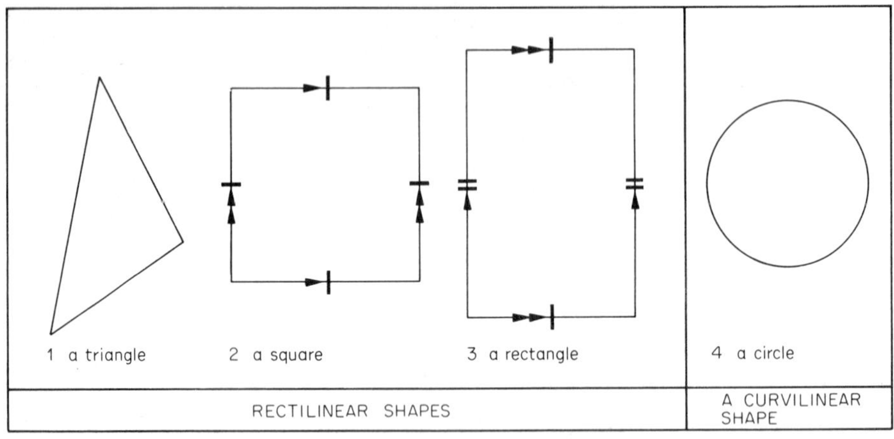

The first shape is a triangle. Triangles have three sides.

The second shape is a square. Squares have four sides. The sides are all equal. The opposite sides are parallel.

The third shape is not a square. It is a rectangle. Rectangles also have four sides but the sides are not all equal. The opposite sides are equal. The opposites sides are equal and parallel.

Triangles, squares and rectangles have straight sides. They are rectilinear shapes. Rectilinear shapes, have straight sides.

The fourth shape is a circle. A circle is not a rectilinear shape. It is a curvilinear shape. Curvilinear shapes have curved sides.

Exercise 1 Some of these statements are *true*. Some of these statements are *not true*. They are *false*. Read the sentences. Are they true or false?

1. The first shape is a triangle.
2. Triangles have four sides.
3. The second shape is a rectangle.
4. A square has four equal sides.
5. A rectangle also has four equal sides.
6. The third shape is not a rectangle.
7. Triangles, rectangles and squares have straight sides.
8. Triangles, rectangles and squares are curvilinear shapes.
9. The fourth shape is not a rectilinear shape.
10. Circles are curvilinear shapes.

Exercise 2 Look at the pictures and complete the sentences.

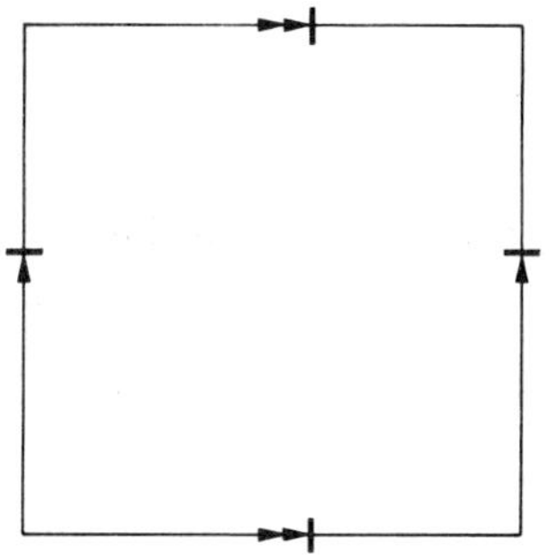

This shape . . . . square.
Squares . . . . rectilinear . . . .
. . . . four equal . . . .
. . . . all equal.
. . . . opposite . . . . equal . . . . parallel.

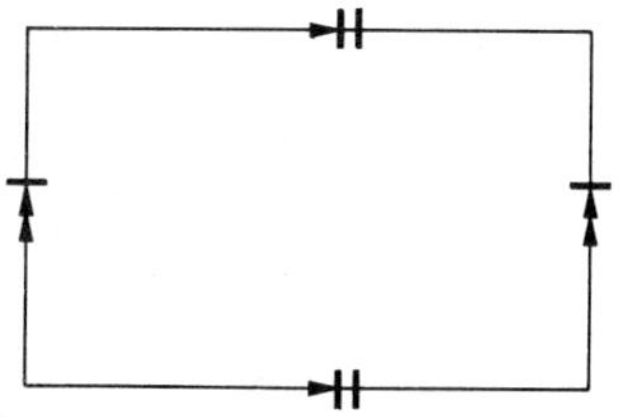

This . . . . square.
It . . . . rectangle.
Rectangles . . . . rectilinear . . . .
. . . . four sides but . . . . all equal.
. . . . opposite . . . . equal . . . . parallel.

Now complete the next three paragraphs with these words.

curvilinear	*opposite*	*rectilinear*
equal	*parallel*	*straight*

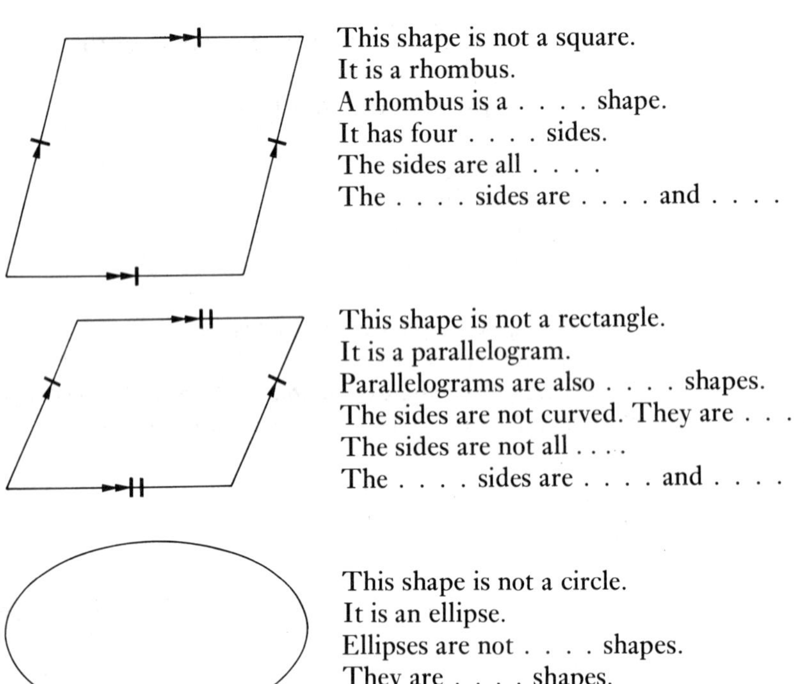

This shape is not a square.
It is a rhombus.
A rhombus is a . . . . shape.
It has four . . . . sides.
The sides are all . . . .
The . . . . sides are . . . . and . . . .

This shape is not a rectangle.
It is a parallelogram.
Parallelograms are also . . . . shapes.
The sides are not curved. They are . . . .
The sides are not all . . . .
The . . . . sides are . . . . and . . . .

This shape is not a circle.
It is an ellipse.
Ellipses are not . . . . shapes.
They are . . . . shapes.

Exercise 3 Complete this table. The first one has been done for you.

SHAPE	NAME OF SHAPE	RECTILINEAR	CURVILINEAR
	a square	✓	✗

SHAPE	NAME OF SHAPE	RECTILINEAR	CURVILINEAR
	a trapezium		
	a trapezoid		

Exercise 4 Look at this example.

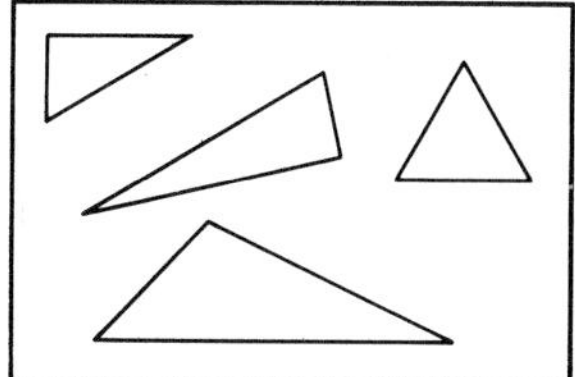

A triangle has three sides.
Triangles have three sides.

These two sentences have the same meaning.

Look at the pictures and complete the sentences in the same way.

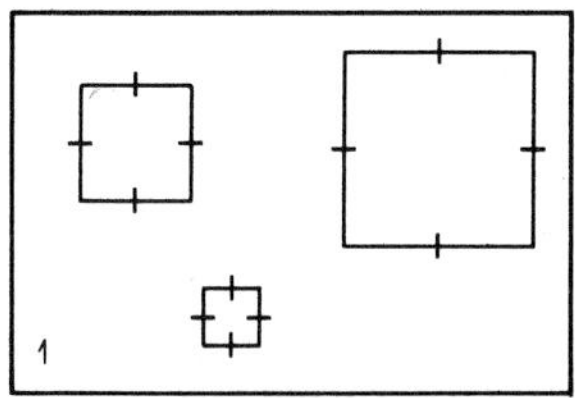

A square . . . . four equal . . . .

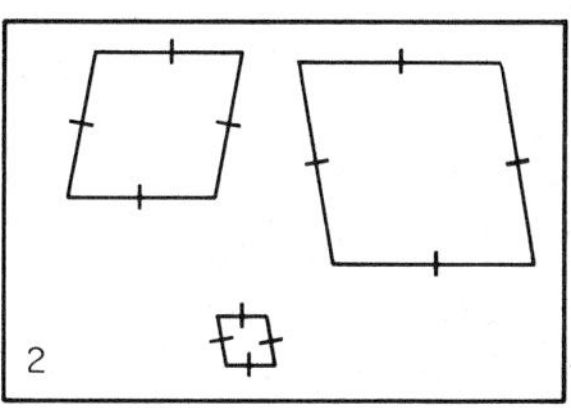

Rhombuses . . . . equal . . . .

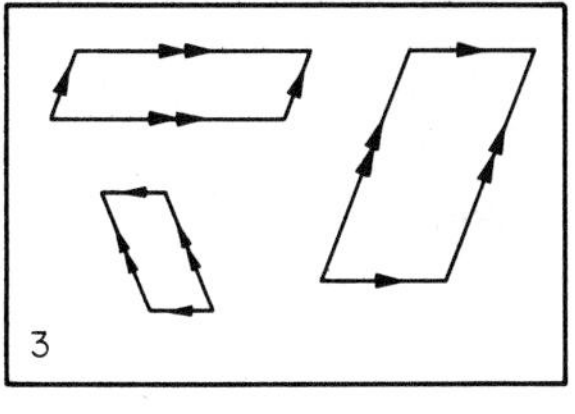

. . . . straight . . . .

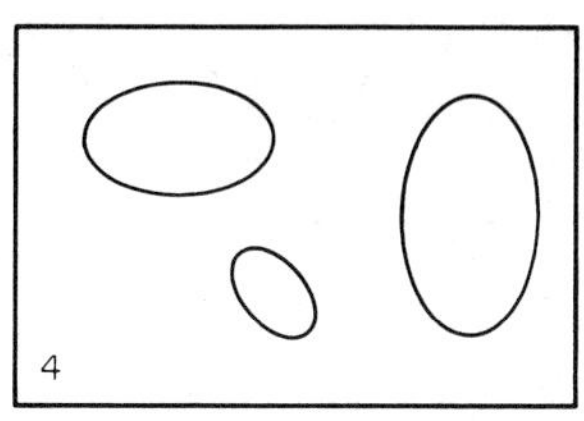

. . . . curved . . . .

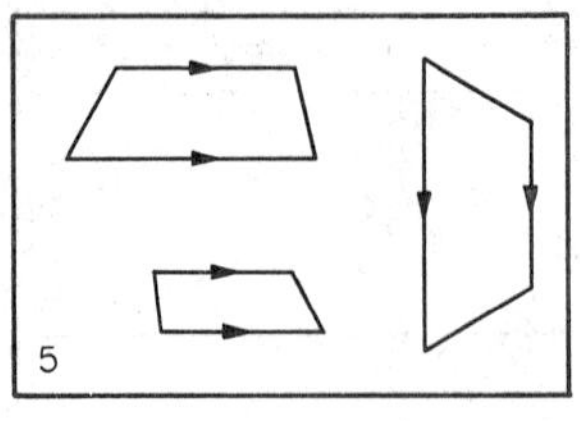

. . . . two parallel . . . .

Now look at this example.

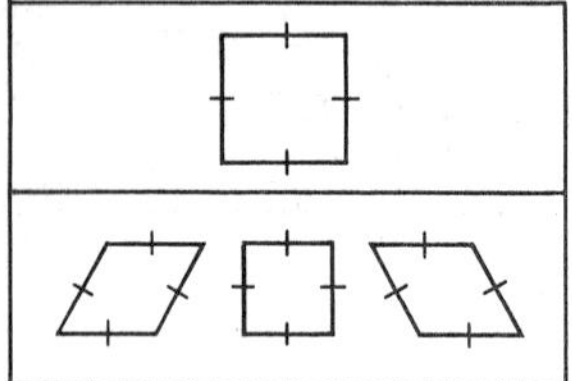

This shape has four equal sides.
These shapes have four equal sides.

Now complete these sentences in the same way.

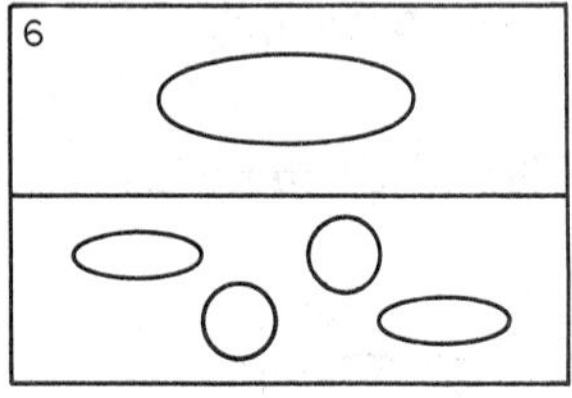

. . . . curved sides.
. . . . curved sides.

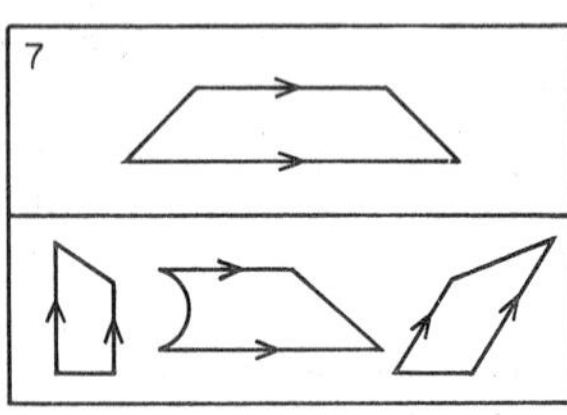

. . . . two parallel . . . .
. . . .

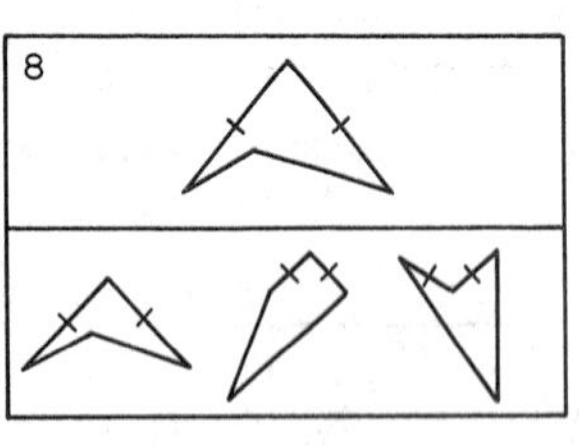

. . . . two equal . . . .
. . . .

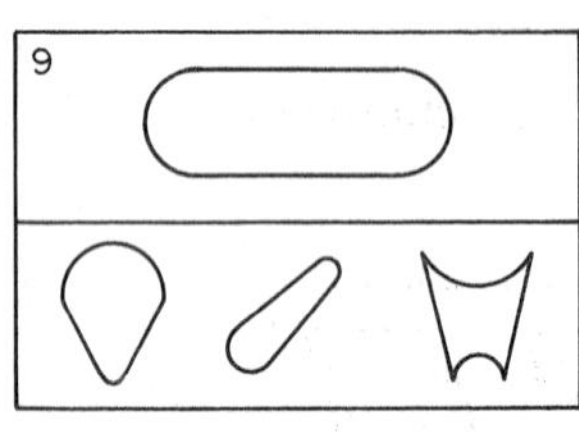

. . . . two . . . .
. . . .

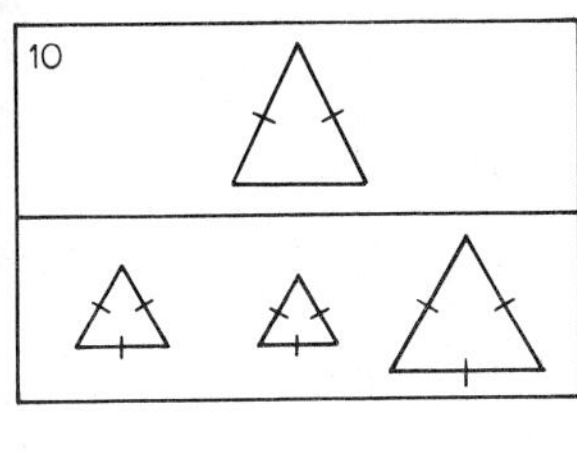

. . . .
. . . .

Exercise 5 Look at this example.

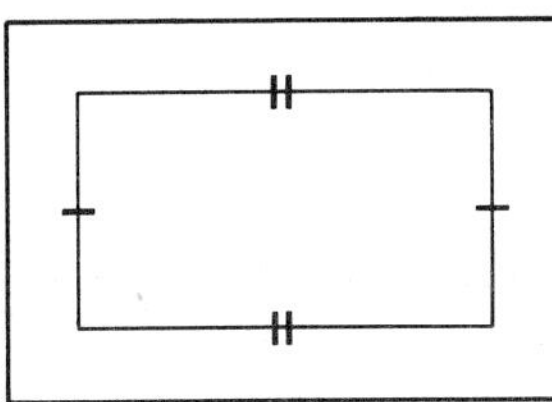

A square has *four sides.*
This shape *also* has *four sides but* it is not a square.

Now complete these sentences in the same way.
Use the words *also* and *but*.

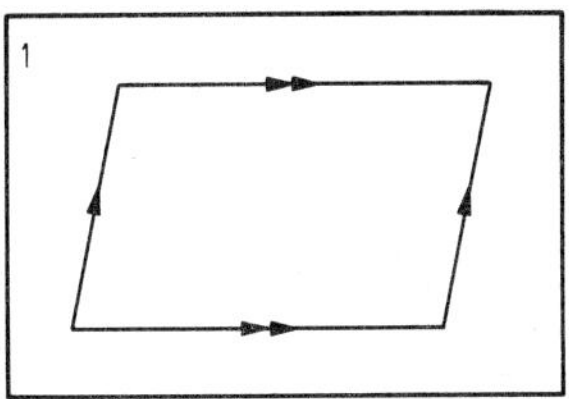

A rectangle has parallel sides.
This . . . . not a rectangle.

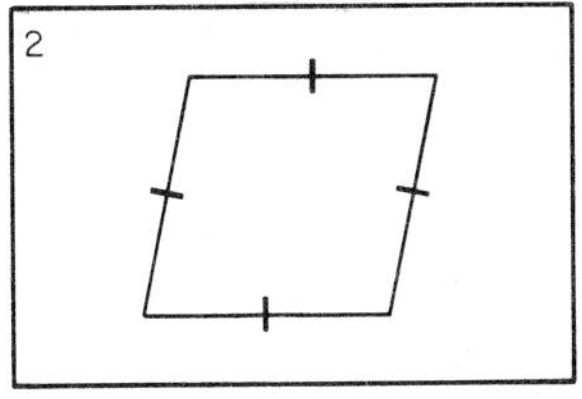

A square has four equal sides.
This . . . . a square.

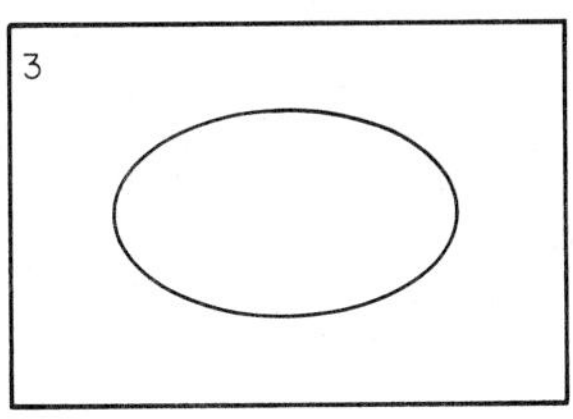

A circle is curvilinear.
This . . . .

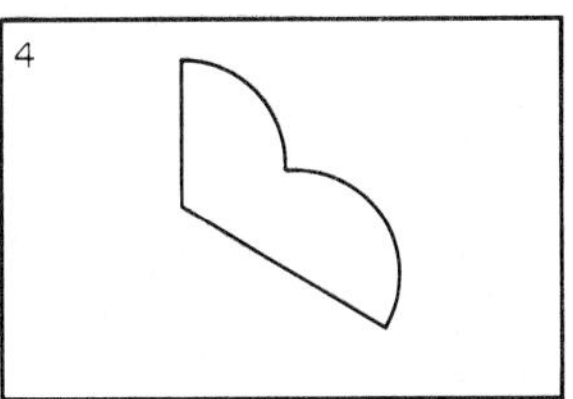

Trapezoids have four sides.
. . . .

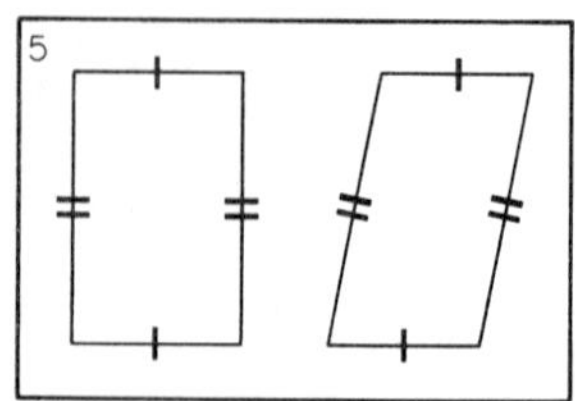

Squares and rhombuses have parallel sides and the sides are all equal.
Rectangles and . . . . equal.

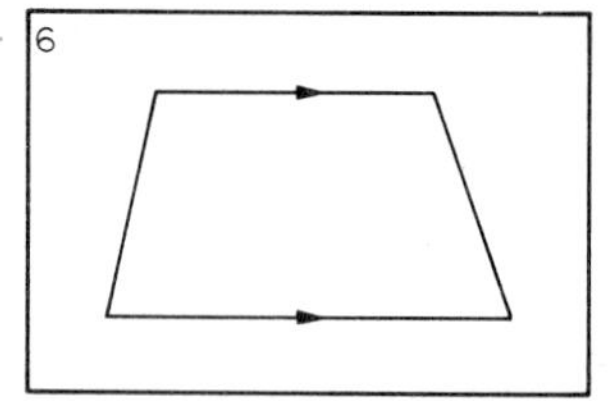

A trapezium has two parallel sides . . . .
. . . . equal.

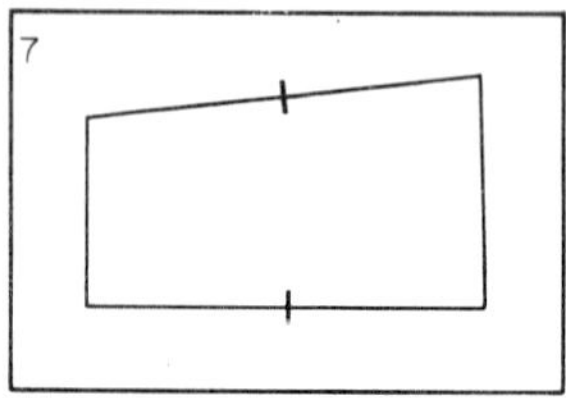

This trapezoid has . . . .
. . . . parallel.

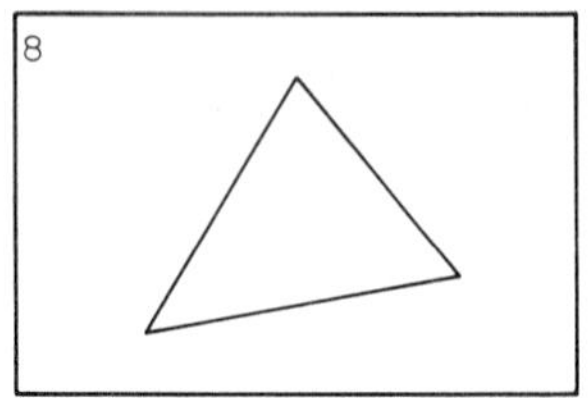

Trapezoids are rectilinear and they have four sides.
Triangles . . . . sides.

a plane
a shape
a side

a triangle
a square
a rectangle
a circle
a rhombus
a parallelogram
an ellipse
a trapezium
a trapezoid

geometrical
equal
opposite
parallel
straight
curved
rectilinear
curvilinear

but

all

SECTION B: ANGLES

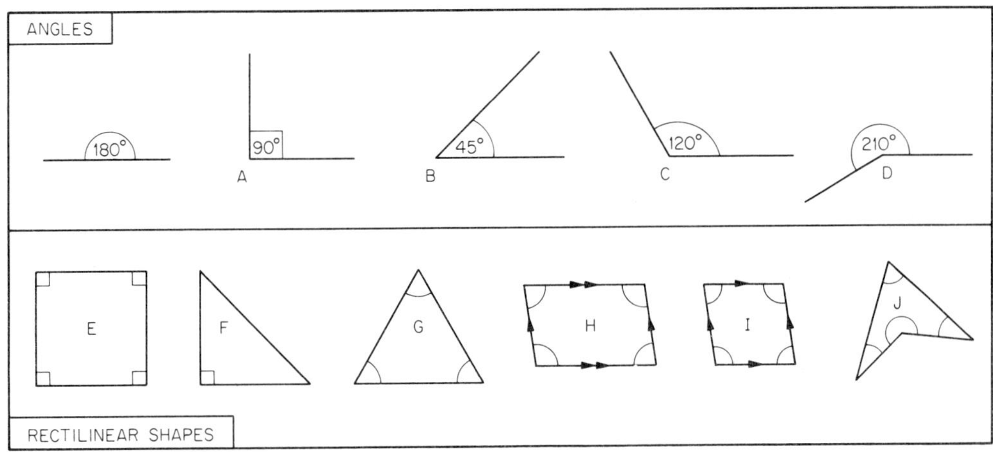

All rectilinear shapes have angles. Curvilinear shapes do not have angles.

A right angle is 90°. Angle A is a right angle. An acute angle is less than 90°. Angle B is an acute angle. It is 45°. An obtuse angle is more than 90° but it is less than 180°. Angle C is an obtuse angle. It is 120°. A reflex angle is more than 180° but less than 360°. Angle D is a reflex angle. It is 210°.

Square E has four right angles. Squares and rectangles always have four right angles. Triangles sometimes have one right angle. Triangle F has a right angle. Triangle G does not have a right angle. It has three acute angles.

Parallelogram H does not have a right angle. The parallelogram and the rhombus do not have right angles. Parallelograms and rhombuses never have right angles. They always have two acute angles and two obtuse angles. Trapezoids sometimes have a reflex angle. Trapezoid J has a reflex angle. It also has one obtuse angle and two acute angles.

Exercise 6 Describe the angles of these shapes.

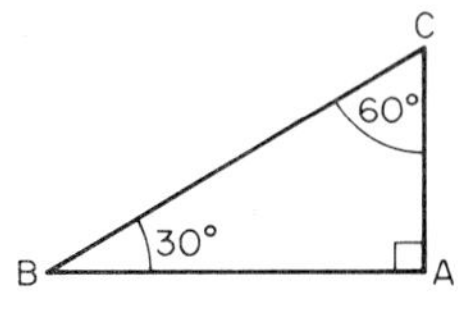

Angle A is . . . . ° It is a . . . . angle.
Angle B is . . . . ° It is . . . . angle.
It is . . . . 90°.
Angle C is . . . . ° It is . . . . angle.
It . . . . 90°.

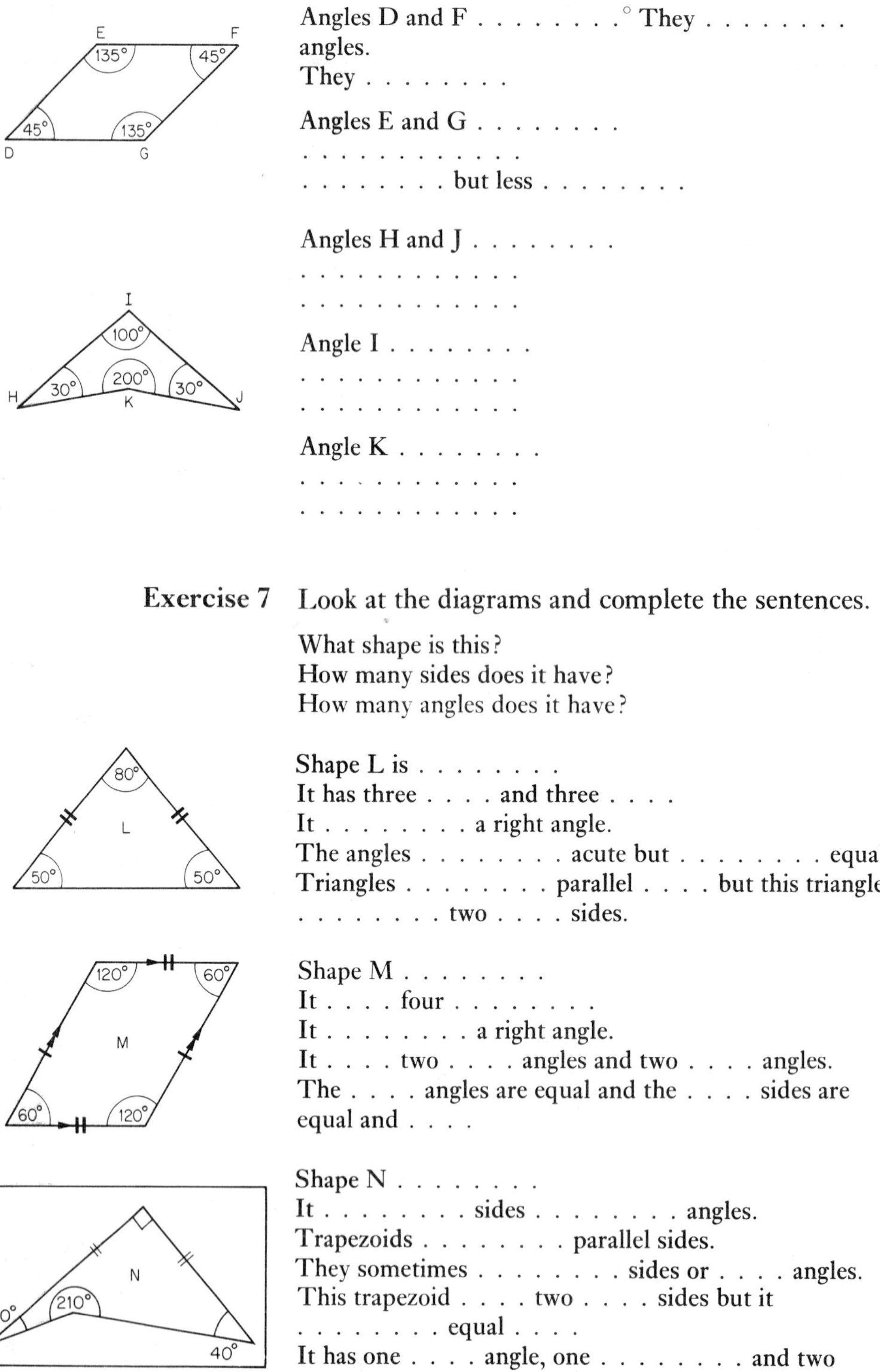

Angles D and F . . . . ° They . . . . angles.
They . . . .

Angles E and G . . . .
. . . .
. . . . but less . . . .

Angles H and J . . . .
. . . .
. . . .

Angle I . . . .
. . . .
. . . .

Angle K . . . .
. . . .
. . . .

Exercise 7 Look at the diagrams and complete the sentences.

What shape is this?
How many sides does it have?
How many angles does it have?

Shape L is . . . .
It has three . . . . and three . . . .
It . . . . a right angle.
The angles . . . . acute but . . . . equal.
Triangles . . . . parallel . . . . but this triangle . . . . two . . . . sides.

Shape M . . . .
It . . . . four . . . .
It . . . . a right angle.
It . . . . two . . . . angles and two . . . . angles.
The . . . . angles are equal and the . . . . sides are equal and . . . .

Shape N . . . .
It . . . . sides . . . . angles.
Trapezoids . . . . parallel sides.
They sometimes . . . . sides or . . . . angles.
This trapezoid . . . . two . . . . sides but it . . . . equal . . . .
It has one . . . . angle, one . . . . and two . . . .

Exercise 8 Look at diagram P.

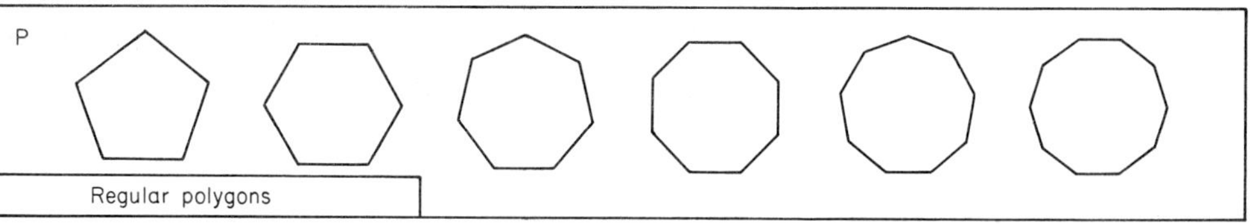

Polygons are rectilinear shapes but they have *more than* four angles.
Complete the table.

SHAPE	NUMBER OF ANGLES	PREFIX	NAME
	5	*penta-*	*pentagon*
			hexagon
		septa-	
			octagon
		nona-	
			decagon

Exercise 9 Look at diagram Q.

Some shapes are *regular*. Some shapes are not regular. Regular shapes have equal angles and equal sides. Squares are regular. Rectangles, parallelograms and rhombuses are not regular. The shapes in diagram P are all regular. The shapes in diagram Q are not regular.

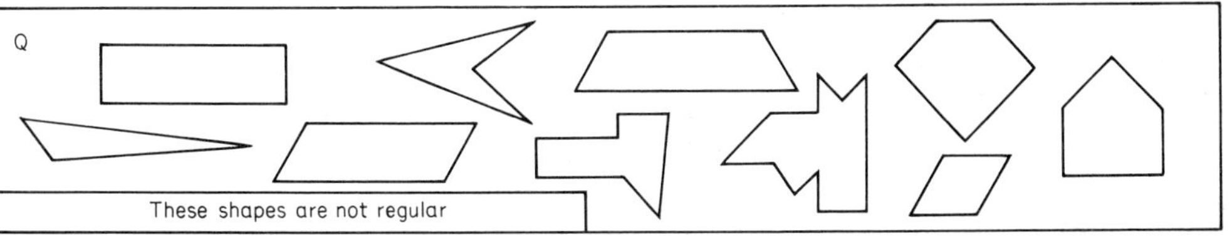

Draw three shapes: a regular *pentagon*, a regular *hexagon* and a regular *octagon*. Label the angles with letters (A, B, C, etc.). Mark the equal sides and the parallel sides. Then answer the questions.

What shape is it?
How many angles does it have?
How many sides does it have?
Are the angles all equal?
Are the sides all equal?
Are the opposite sides parallel?

1. Complete the sentences: Regular pentagons have . . . .

2. Complete the sentences: Regular hexagons . . . .

3. Complete the sentences: Regular octagons . . . .

Exercise 10 Make sentences from this table.

Squares Rectangles Rhombuses Parallelograms Trapezoids Triangles	always sometimes never	have	a two three four	right acute obtuse reflex	angle(s).

a right angle
a degree (°)

a polygon
a pentagon
a hexagon
a septagon
an octagon
a nonagon
a decagon
a diagram

acute
obtuse
reflex
regular

draw
label
mark

always
sometimes
never

more than
less than

how many . . . . ?

SECTION C: SOLID SHAPES

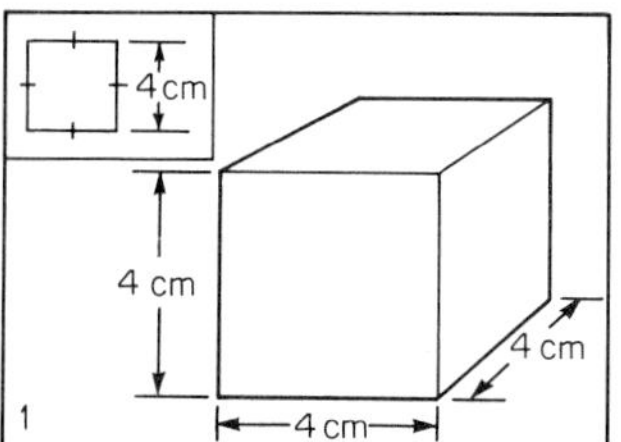

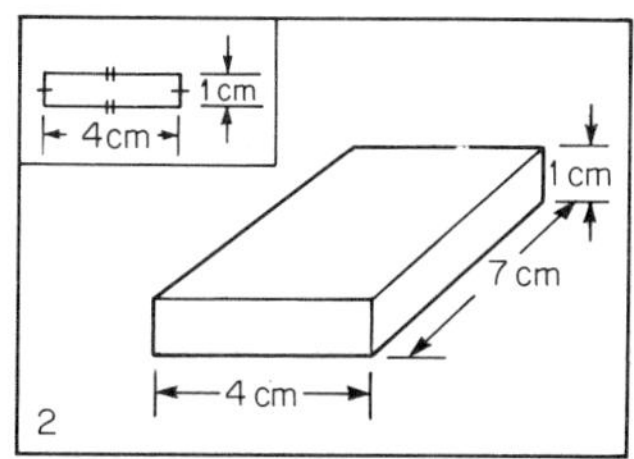

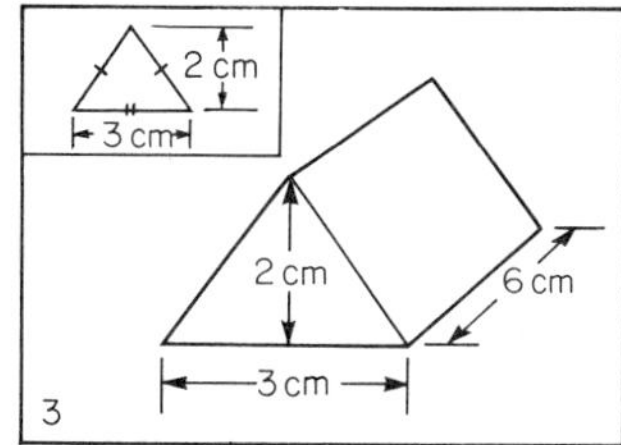

A plane shape has two dimensions. It has a length and a height. Solid shapes have three dimensions. They have a length, a height and a width.

The first shape is a cube. A cube has six faces. All the faces are the same shape. The length, the height and the width are always the same. This cube is 4 cm long, 4 cm high and 4 cm wide.

The next solid is a rectangular prism. Rectangular prisms also have six faces. All the faces are rectangular, but they are not always the same shape. This prism is 7 cm long, 1 cm high and 4 cm wide.

The last solid is another prism but it is not rectangular. It is a triangular prism. Triangular prisms have five faces. They have two triangular faces and three rectangular faces. The triangular faces are always the same shape. The rectangular faces are not always the same shape. This triangular prism is regular and all the rectangular faces are the same. The prism is 6 cm long, 3 cm wide and 2 cm high.

Exercise 11 This solid is a regular *pentagonal* prism.

How many faces does it have?
Are they all the same shape?
What are the dimensions?

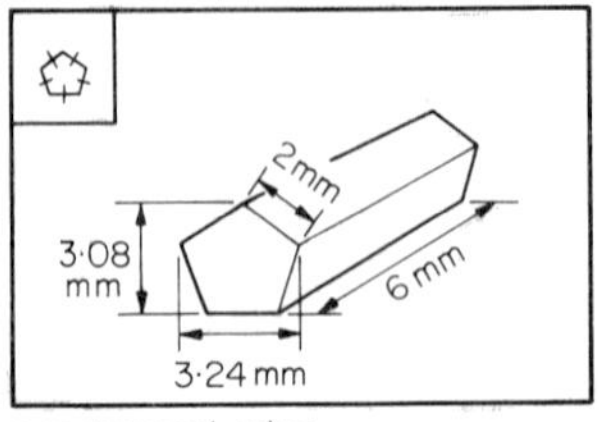

a pentagonal prism

Look at this diagram and complete the paragraph.

A pentagonal prism has . . . . faces. The faces . . . . same shape. It has . . . . rectangular . . . . and . . . . pentagonal faces. The rectangular . . . . are all the same . . . . They are . . . . long and . . . . high. The prism is . . . . long, 3.24 mm . . . . and 3.08 mm . . . .

Now look at these two diagrams and write two paragraphs.

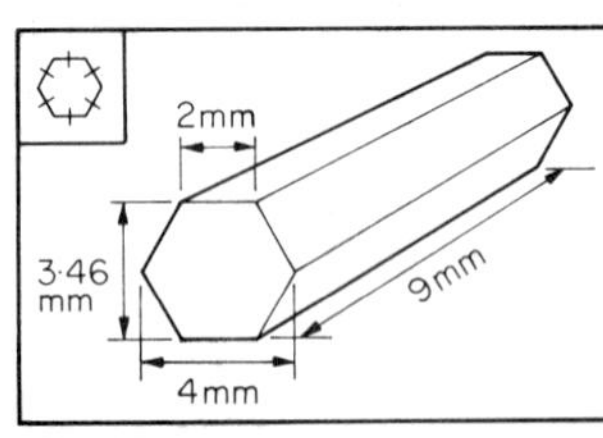

a hexagonal prism

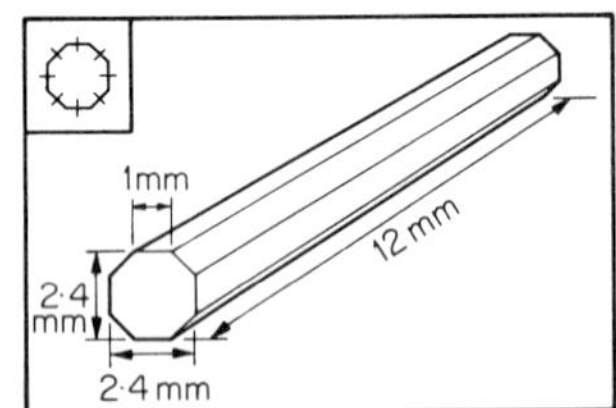

an octagonal prism

Exercise 12 Complete this table. Some of it has been done for you.

SHAPE	NAME OF SHAPE	ADJECTIVE
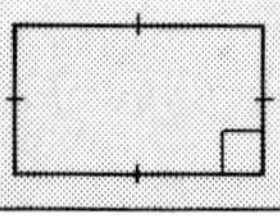	*a square*	*square*
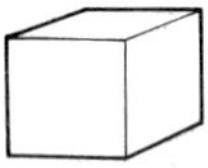		*cubic*
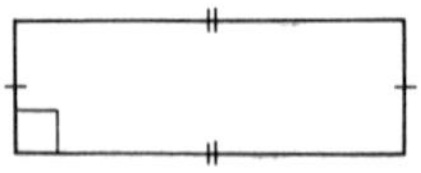		

SHAPE	NAME OF SHAPE	ADJECTIVE
(pentagon)		
(hexagon)		
(octagon)		
(circle)		*circular*
(semicircle)	*a semicircle*	
(ellipse)		*elliptical (or oval)*

Exercise 13 These metal shapes are used in engineering.

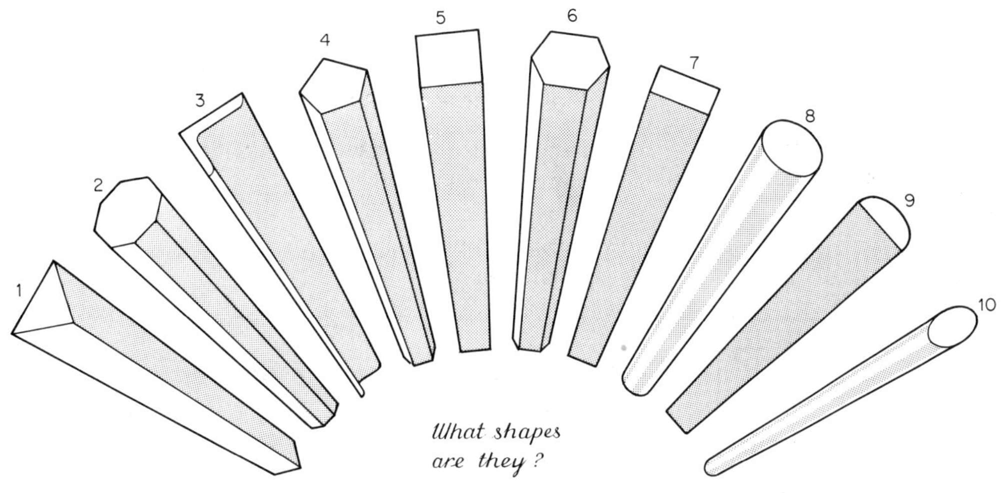

Exercise 14 Look at the example.

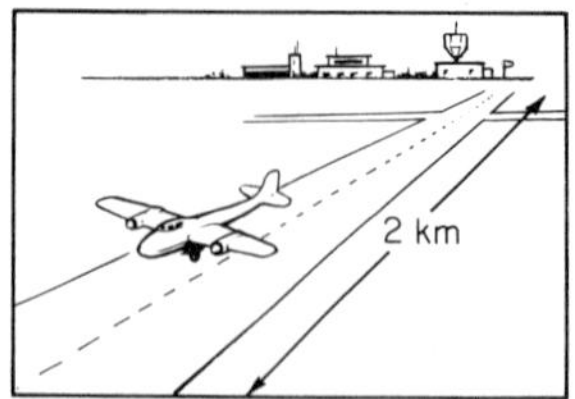

How long is this runway?
It is 2 km *long.*
It has a length of 2 km.

These two sentences have the same meaning.

Describe these dimensions in the same way.

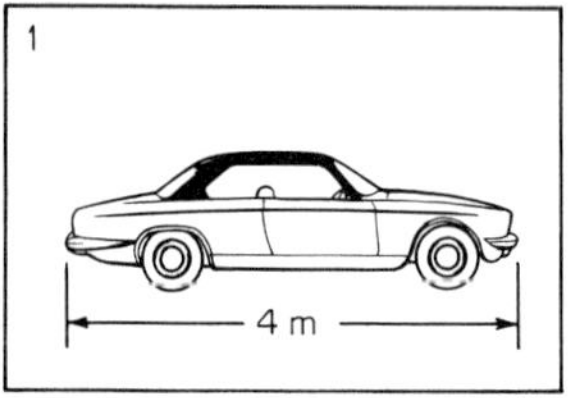

How long is this car?

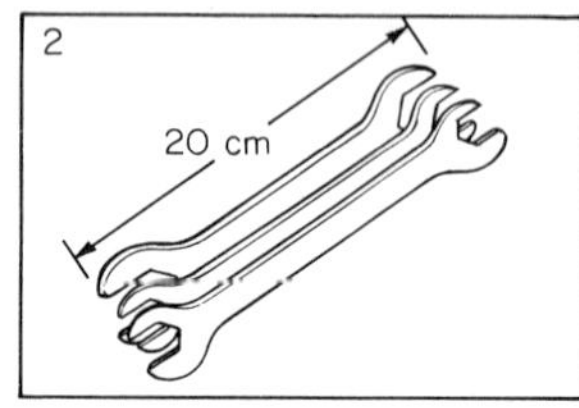

How long are these spanners?

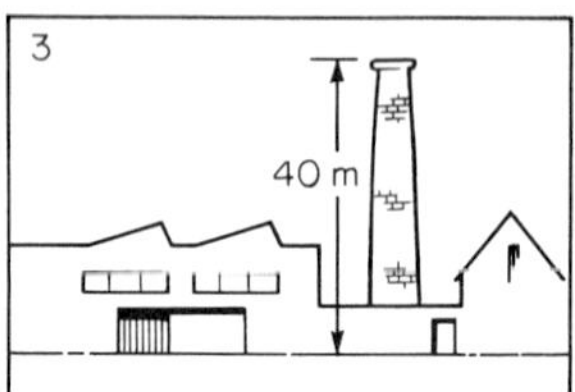

How high is this chimney?

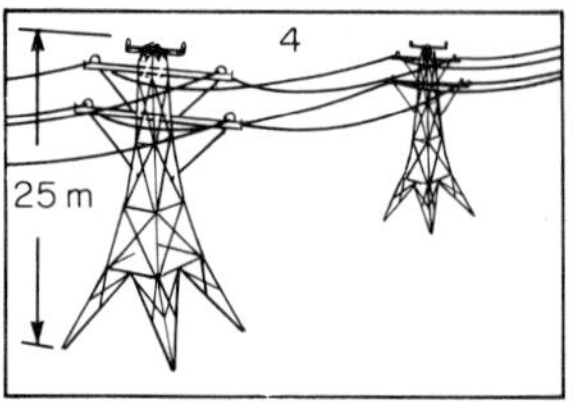

How high are these pylons?

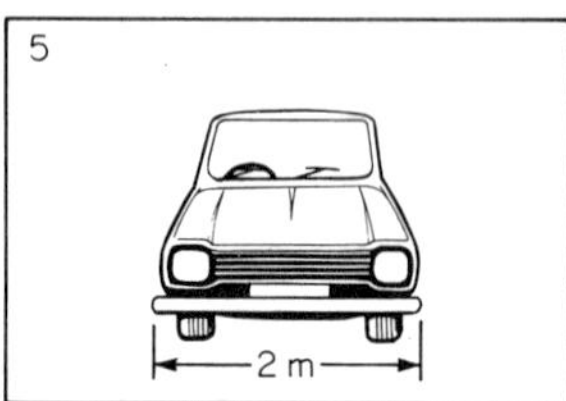

How wide is this car?

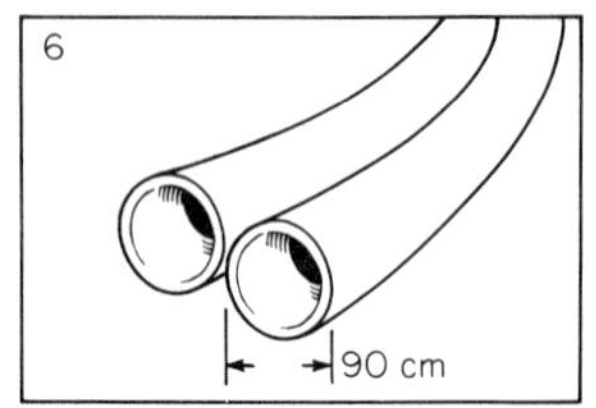

How wide are these pipes?

Now look at these examples.

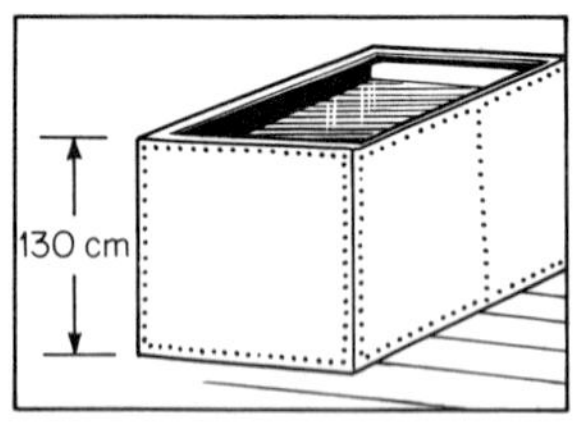

A container does not always have a height.
It sometimes has a *depth.*

How *deep* is this tank?
It is 130 cm *deep.*
It has a depth of 130 cm.

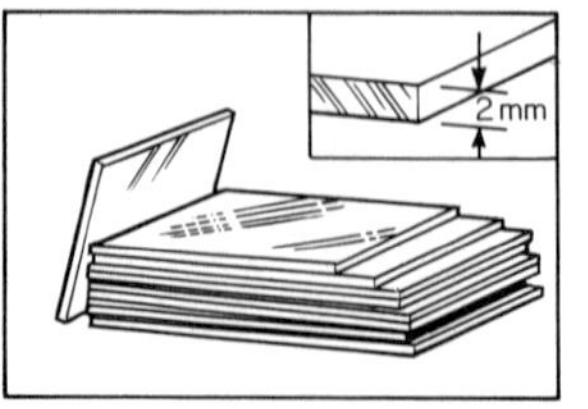

Materials do not always have a height.
They sometimes have a *thickness.*

How *thick* is this glass?
It is 2 mm *thick.*
It has a thickness of 2 mm.

Describe these dimensions in the same way.

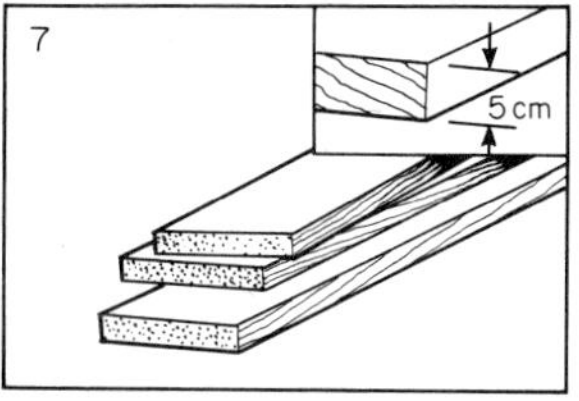

How thick is this wood?

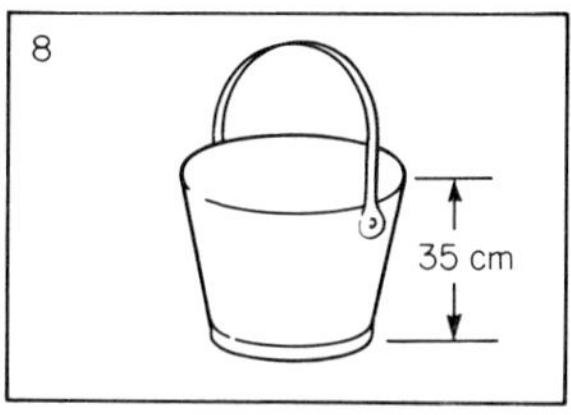

How deep is this bucket?

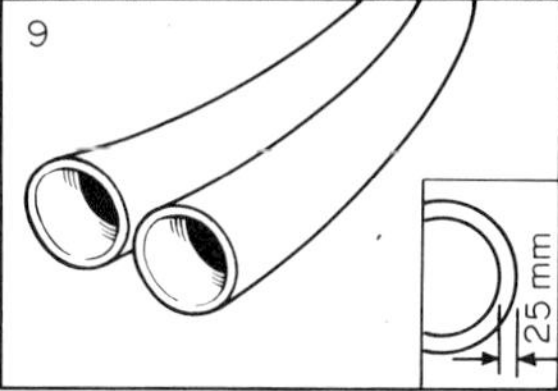

How thick are these pipes?

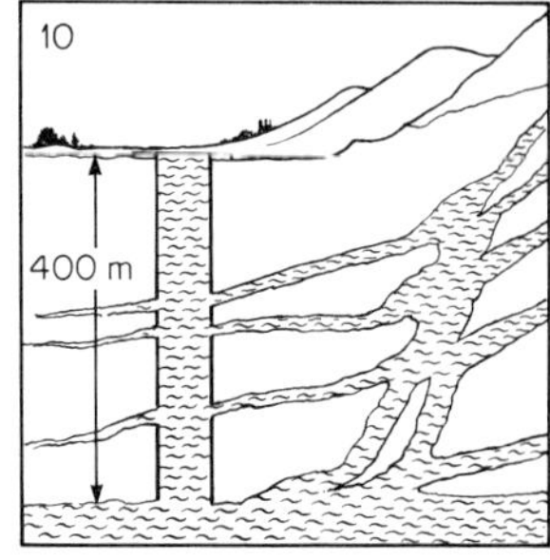

How deep is this well?

Exercise 15 Listen to the teacher. Complete the diagrams and the sentences.

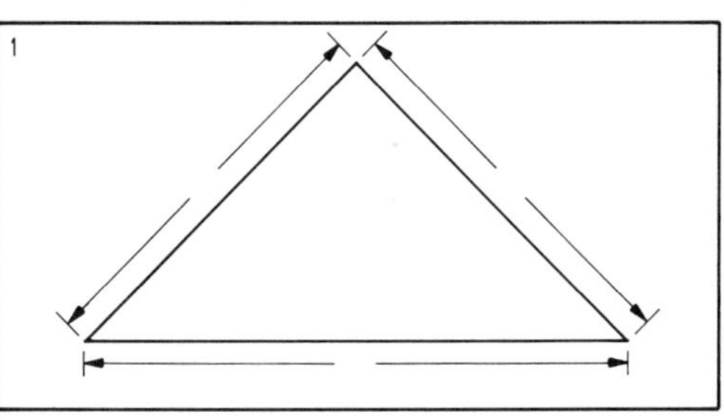

1. Angle A is . . . .°
2. Angle B
3. Angle C
4. Side AB
5. Side BC
6. Side AC

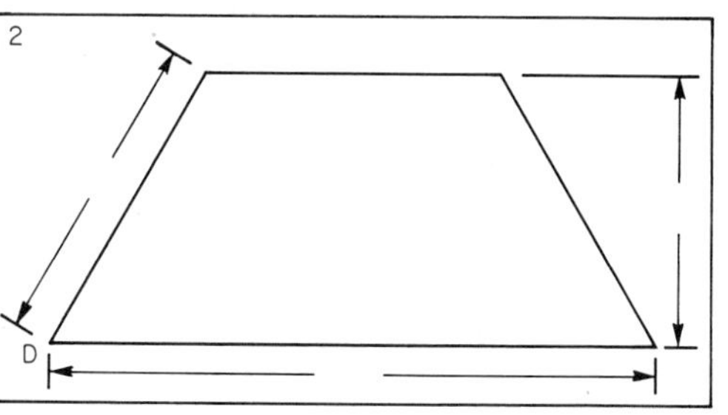

1. Side EF
2. Side DG
3. Sides DE and FG
4. It has a height . . . .
5. Angles D and G
6. Angles E and F

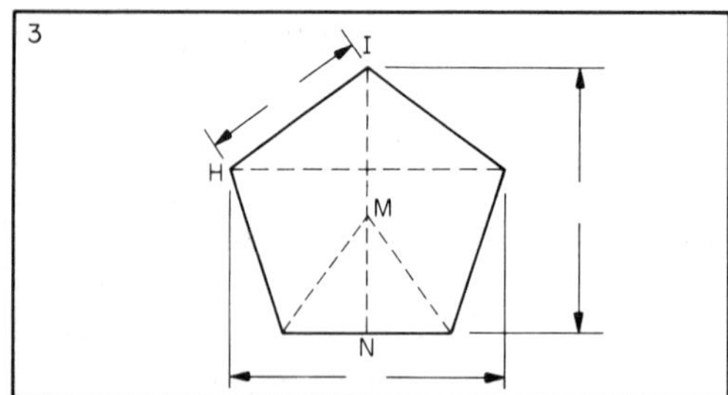

1. All the angles . . . .
2. All the sides . . . .
3. It has a length . . . .
4. It has a height . . . .

Triangle LMK

5. Angles MLN and MKN . . . .
6. Angle LMK . . . .

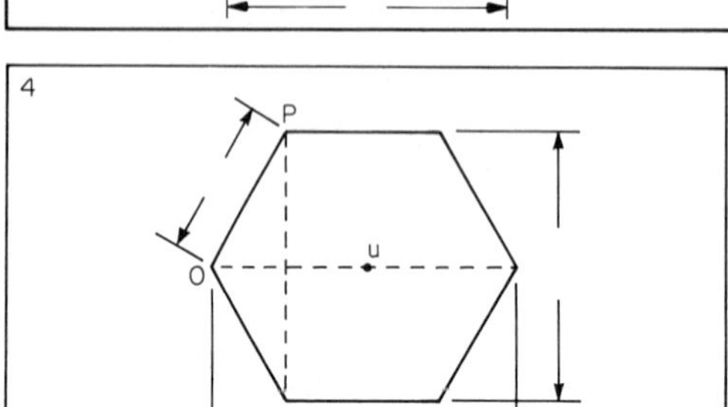

1. All the angles . . . .
2. All the sides . . . .
3. It is . . . . long
4. It is . . . . high

Triangle TUS

5. All the sides . . . .
6. All the angles . . . .

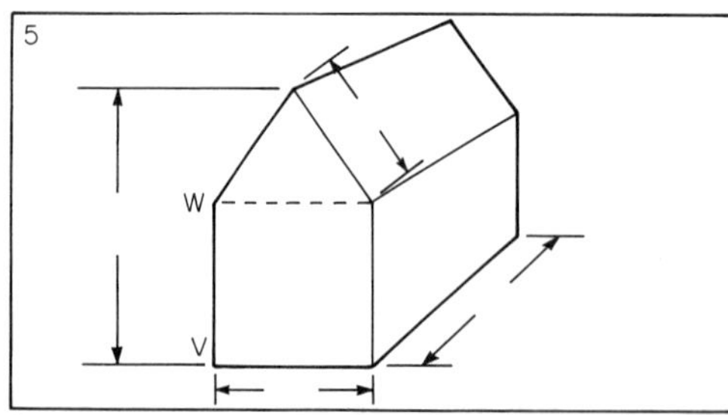

1. All the sides . . . .
2. Angle X . . . .
3. Angles W and Y . . . .
4. Angles V and Z . . . .
5. Shape VWYZ . . . .
6. Shape WXY . . . .

LANGUAGE NOTE 11

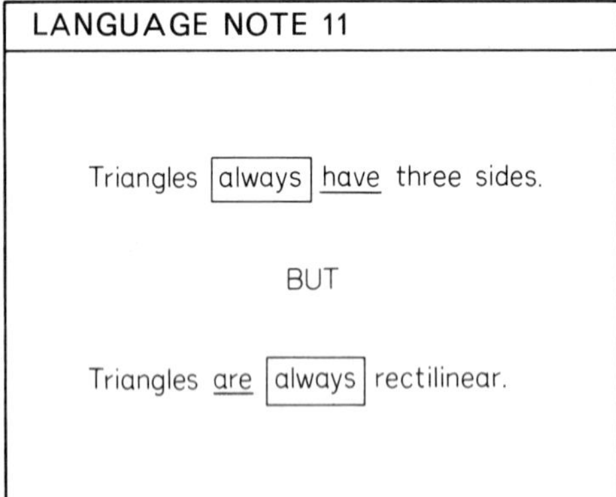

Triangles always have three sides.
BUT
Triangles are always rectilinear.

LANGUAGE NOTE 12

a face
a solid
a cube
a prism
a semicircle

a millimetre (mm)
a centimetre (cm)
a metre (m)
a kilometre (km)

a length (long)
a height (high)
a width (wide)
a depth (deep)
a thickness (thick)

a runway
a car
a chimney
a pylon
a pipe
a well

square
cubic
rectangular
triangular
angular
pentagonal
hexagonal
octagonal
(semi) circular
elliptical (oval)

next
last

same

how . . . . ?

UNIT THREE Motor Vehicles

SECTION A: ENGINES (i)

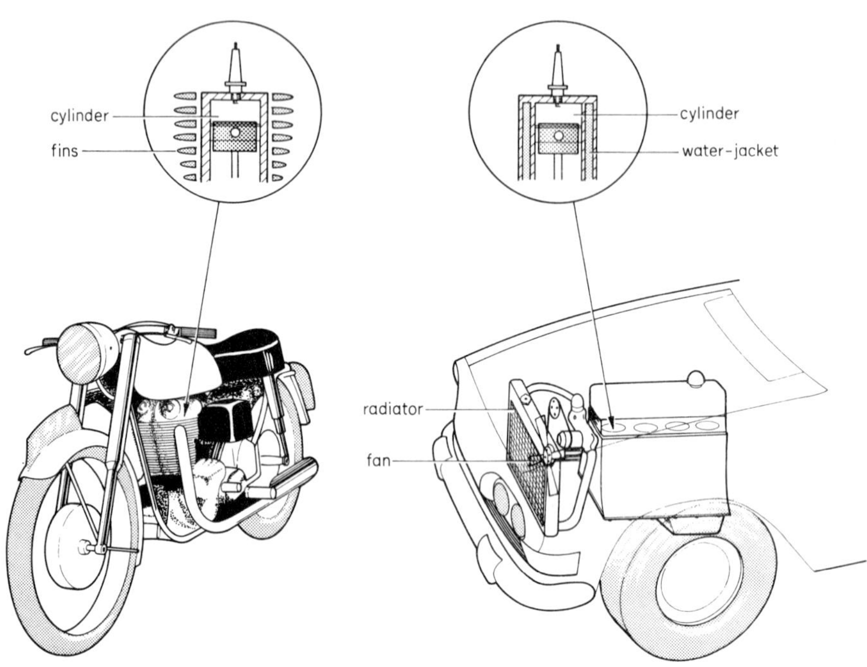

There is a motorcycle in the first picture. There is a car in the second picture. The motorcycle has an air-cooled engine. There is a water-cooled engine in the car.

The motorcycle engine has fins around the cylinders. Air-cooled engines always have these fins. There is a water jacket around the cylinders in the car engine. Water-cooled engines always have a water jacket around the cylinders.

There is a fan at the front of the car engine. Water-cooled engines always have a fan at the front. There is sometimes a fan at the front of air-cooled engines.

Water-cooled engines always have a radiator. There is no radiator on the motorcycle. Air-cooled engines never have a radiator.

The engine is at the front of the car. Car engines are not always at the front. They are sometimes at the back or sometimes in the middle.

Exercise 1 Are these statements true or false?

1. The car has an air-cooled engine.
2. A water-cooled engine has fins around the cylinders.
3. The car engine has a water jacket around the cylinders.
4. There is a fan on the motorcycle engine.
5. There is always a fan on water-cooled engines.
6. Air-cooled engines sometimes have a fan.
7. There is a radiator on the motorcycle.
8. Cars sometimes have an engine at the back or in the middle.

Exercise 2

Look at the example.

This motorcycle *has* an air-cooled engine.
There is an air-cooled engine *in* this motorcycle.

These two sentences have the same meaning.

Look at these pictures and complete the sentences. Use the words *in*, *on* or *around*.

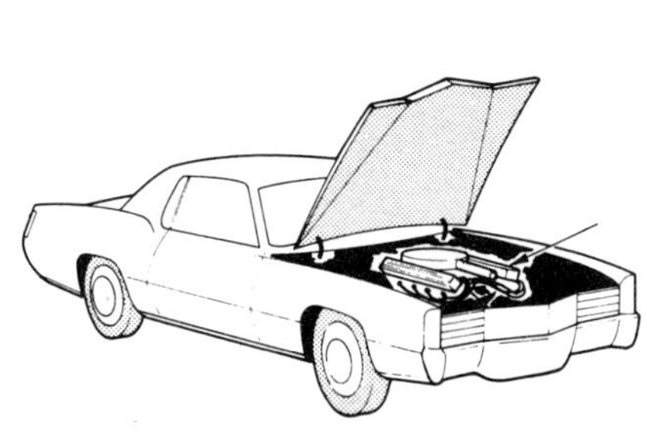

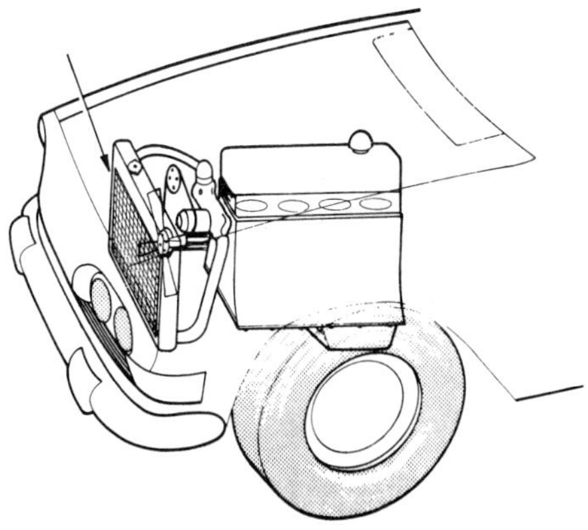

1. This car has a water-cooled engine.
   There is . . . . car.

2. Water-cooled engines have a radiator.
   There . . . . engines.

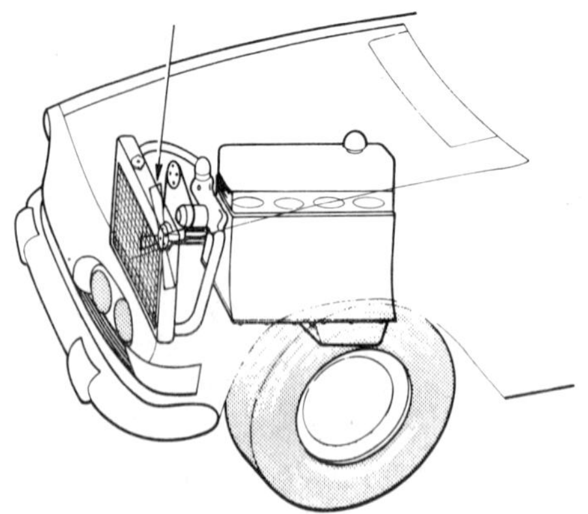

3. The car engine has a fan.
. . . .

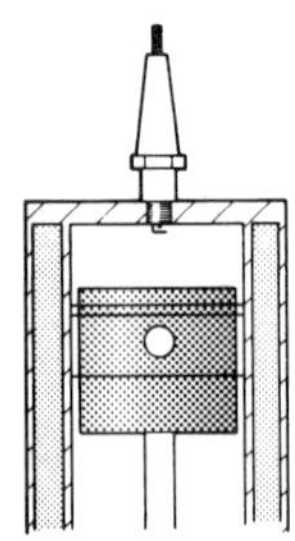

4. This cylinder has a water jacket.
. . . .

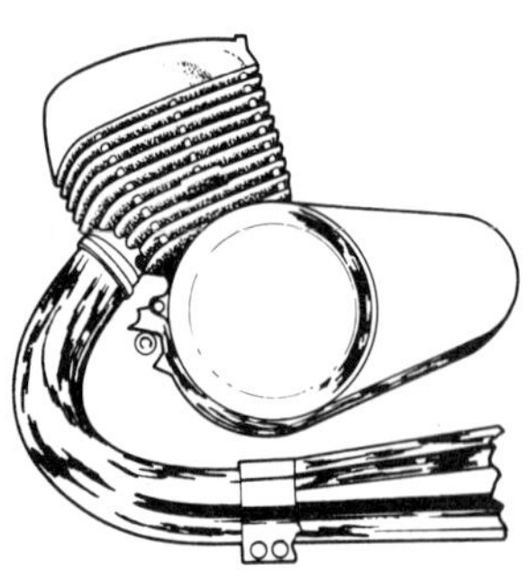

5. An air-cooled engine has no water jacket. . . . .

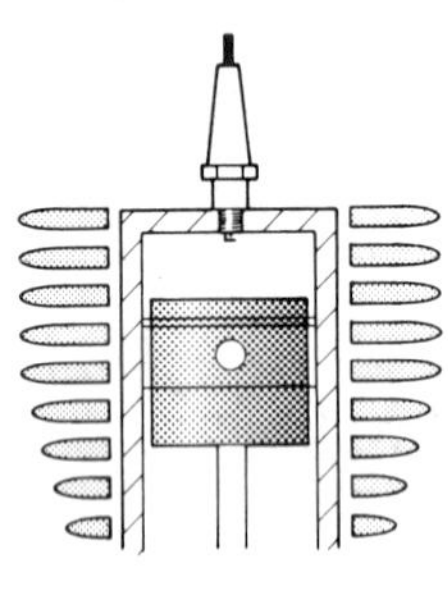

6. This cylinder has no water jacket.
. . . .

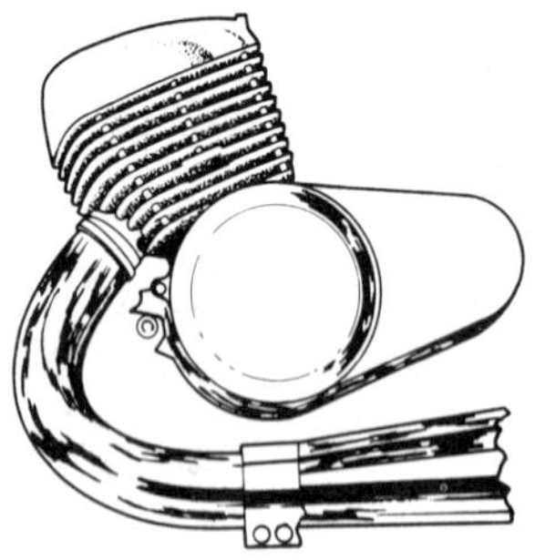

7. This motorcycle has no radiator.
. . . .

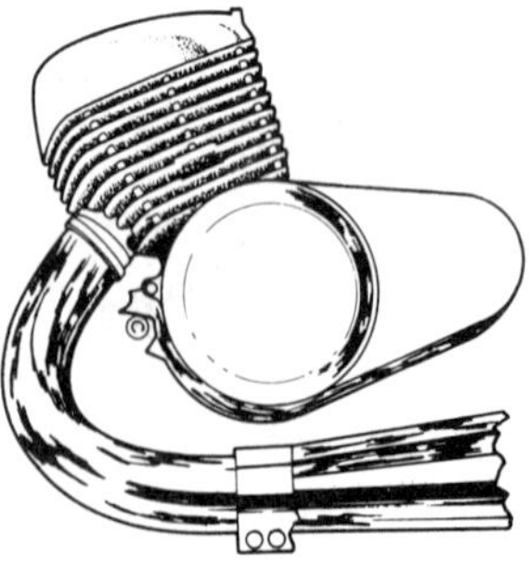

8. The engine has no fan.
. . . .

Exercise 3 Look at the examples.

There *is always* a fan on a water-cooled engine.
A water-cooled engine *always has* a radiator.

Complete these sentences with *always*, *never* or *sometimes*.

1. A water-cooled engine has a water jacket around the cylinders.
2. An air-cooled engine has a water jacket around the cylinders.
3. There is a radiator on an air-cooled engine.
4. There is a fan on an air-cooled engine.
5. Cars have the engine at the back.
6. Air-cooled engines have fins around the cylinders.
7. There is a reflex angle in a trapezoid.
8. There is a right angle in a parallelogram.

Exercise 4 Look at the example.

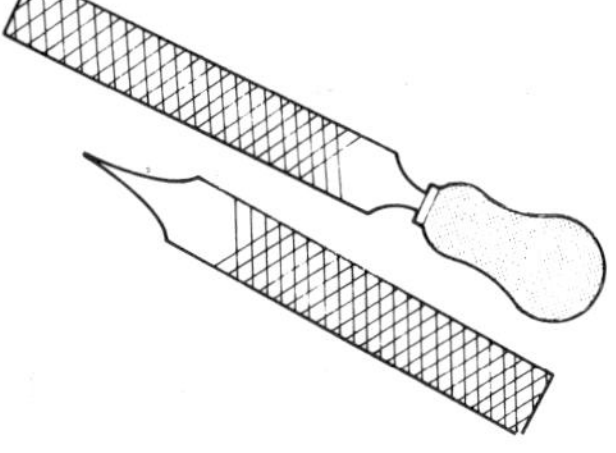

There is a handle on *one* file.

There is no handle on *the other*.

Make sentences from these pictures in the same way. Use *on* or *in*.

1

2
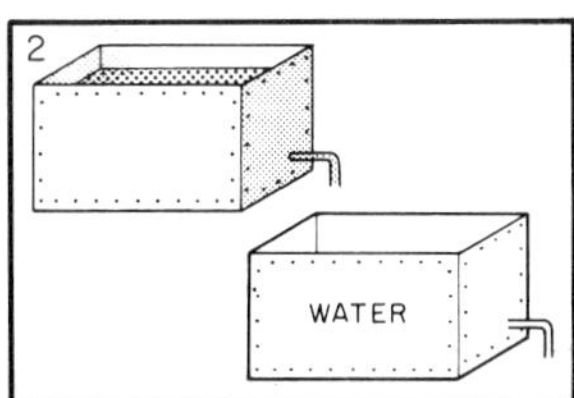

3
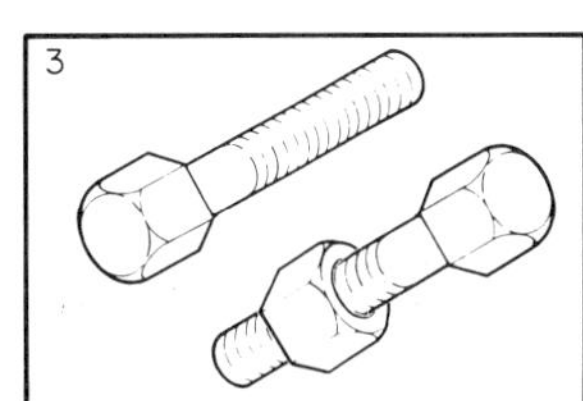

4
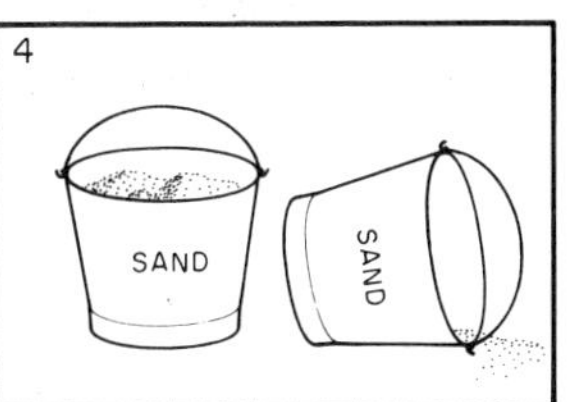

5
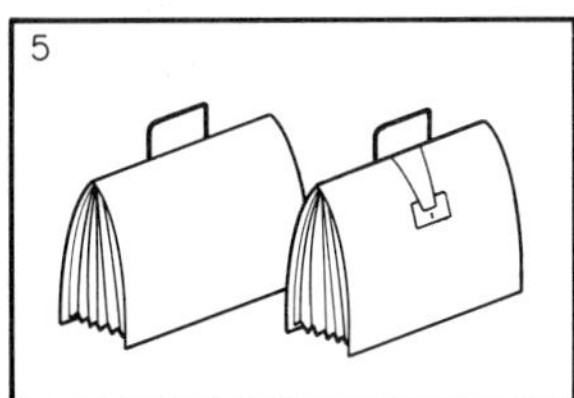

6
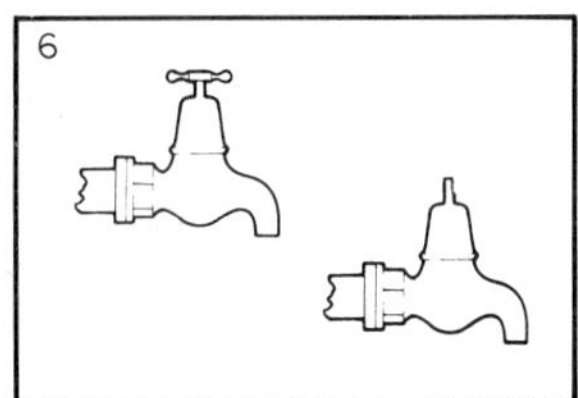

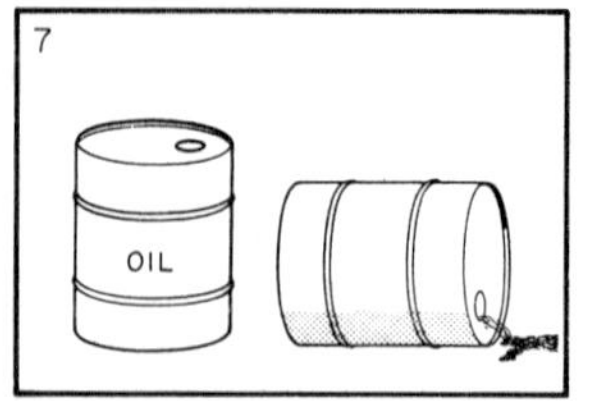

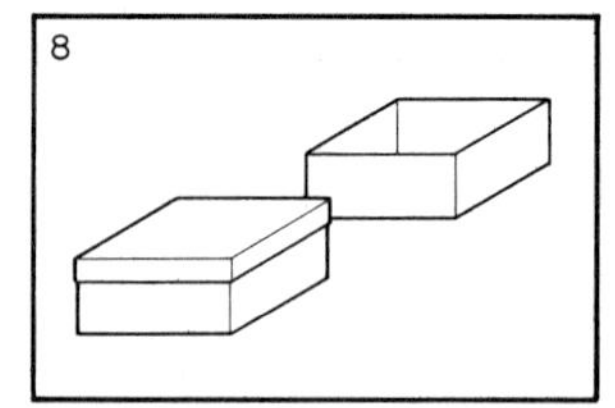

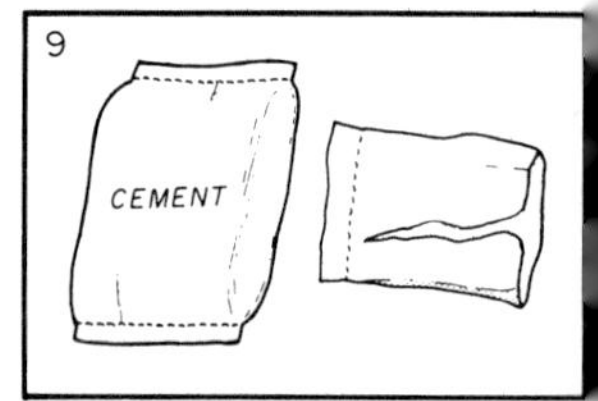

Exercise 5 Look at the example.

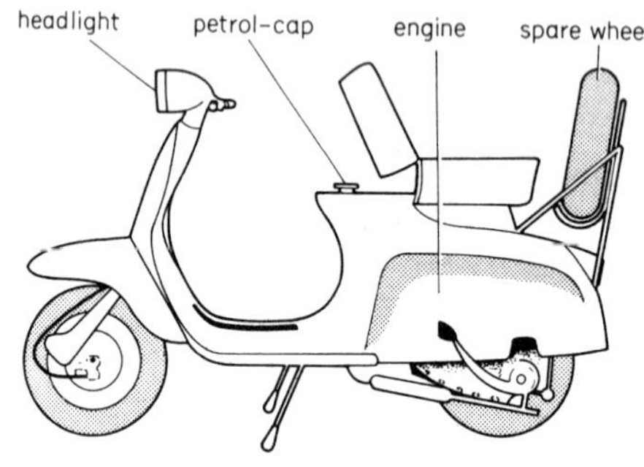

The engine is *at the side of* the scooter.
It has one headlight *at the front.*
The sparc wheel is *at the back.*
There is a petrol cap *in the middle.*

Complete these sentences in the same way.

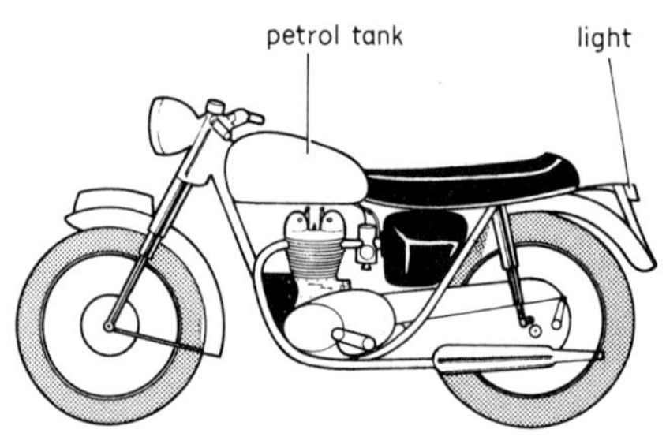

The engine is . . . . the motorcycle.
There is one headlight . . . .
There is another light . . . .
The petrol tank is . . . .

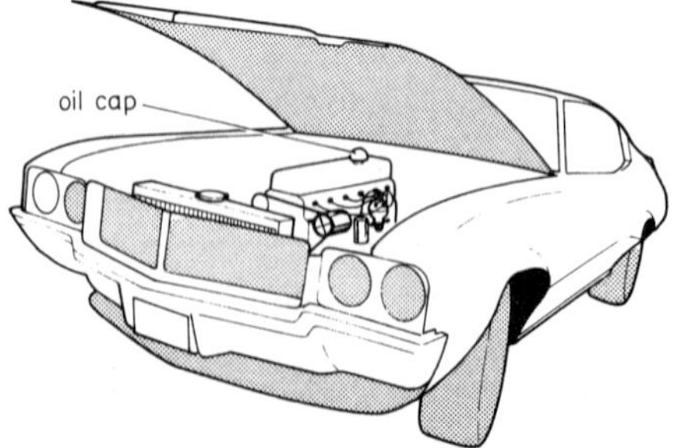

The car has four headlights . . . .
The engine is . . . .
The radiator . . . . of the engine.
There is an oil cap . . . . engine.

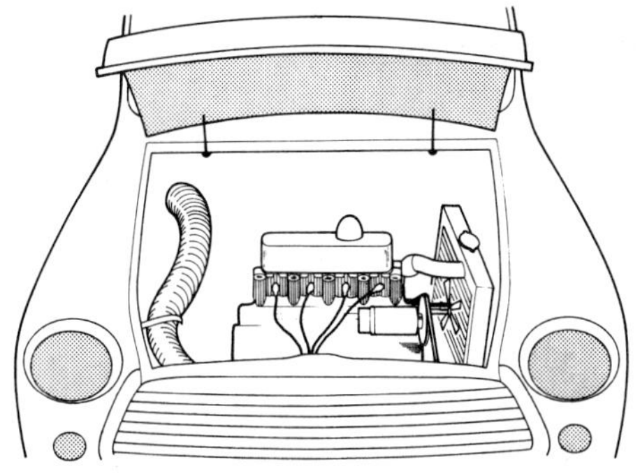

This car has the engine . . . .
The radiator . . . .
There is a fan . . . .
There is a cap . . . .

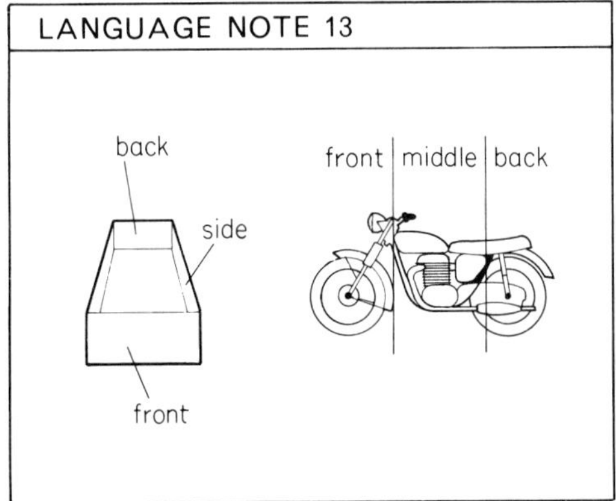

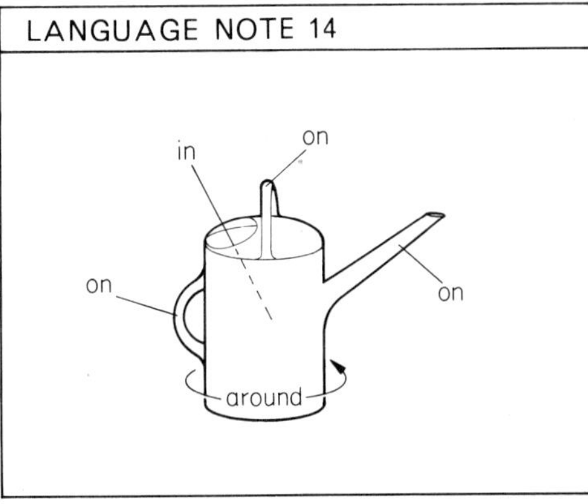

a motor vehicle
an engine
a motorcycle
a scooter
a fin
a cylinder
a jacket
a fan
a radiator
a (head)light
a wheel

air-cooled
water-cooled
spare

other

on
in
around
at the front (of)
at the back (of)
at the side (of)
in the middle (of)

SECTION B: ENGINES (ii)

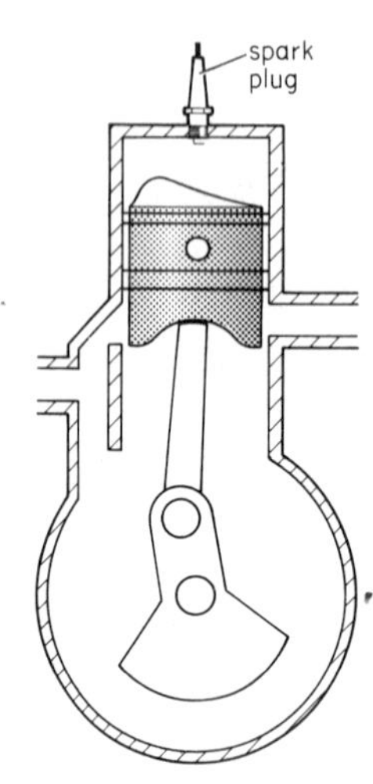

a 2-stroke petrol engine

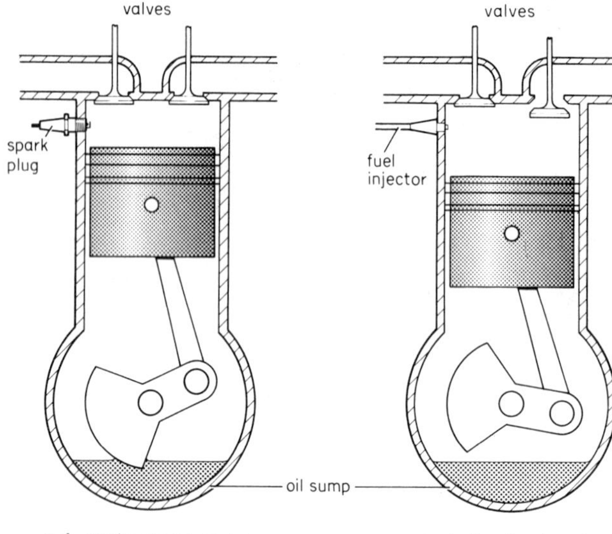

a 4-stroke petrol engine

a 4-stroke diesel engine

All motor vehicles have an engine. There are two types of engine. There are petrol engines and there are diesel engines. There are three engines in the diagram. There is a 2-stroke petrol engine on the left. There is a 4-stroke petrol engine in the middle and there is a 4-stroke diesel engine on the right.

There are spark plugs in both the petrol engines. There are spark plugs in all petrol engines. Diesel engines do not have spark plugs. They have fuel injectors.

There are valves in both the 4-stroke engines. There are always valves in 4-stroke engines. There are no valves in the 2-stroke petrol engine. A 2-stroke petrol engine never has valves.

There is no oil sump in the 2-stroke engine. There is oil in the fuel. Both the 4-stroke engines have an oil sump. There is no oil in the fuel.

Exercise 6 Complete this table from the text:

	oil sump	valves	fuel injector	spark plug
2-stroke petrol engine	X			
4-stroke petrol engine				
4-stroke diesel engine		√		

Now look at these examples.

There is no oil sump in the 2-stroke petrol engine.
There are valves in the 4-stroke diesel engine.

Make ten sentences from the table in the same way.

Exercise 7 Look at this example. There is a 2-stroke *diesel* engine in the diagram.

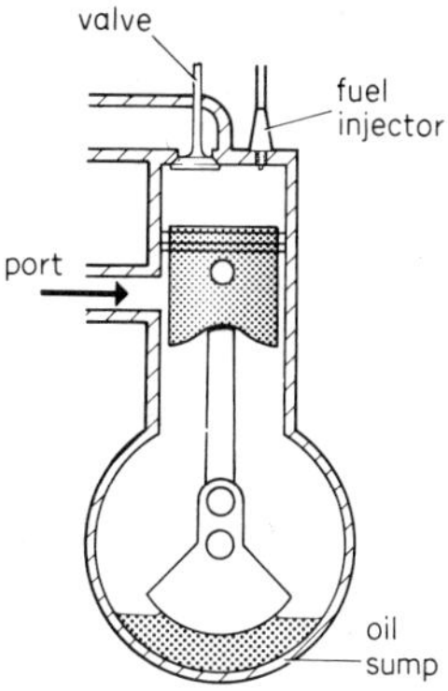

a) *There is* a fuel injector in a 2-stroke diesel engine.

b) A 2-stroke diesel engine *has* a fuel injector.

These two sentences have the same meaning.

Complete these sentences in the same way.

1. There is one valve in this engine.
   This . . . . valve.

2. There is an oil sump in a 2-stroke diesel engine.
   A . . . .

3. There is a fuel injector on all diesel engines.
   . . . .

4. This engine has one port.
   There . . . .

5. This engine has no spark plugs.
   . . . .

6. Diesel engines have no spark plugs.
   . . . .

Exercise 8 Look at the picture.

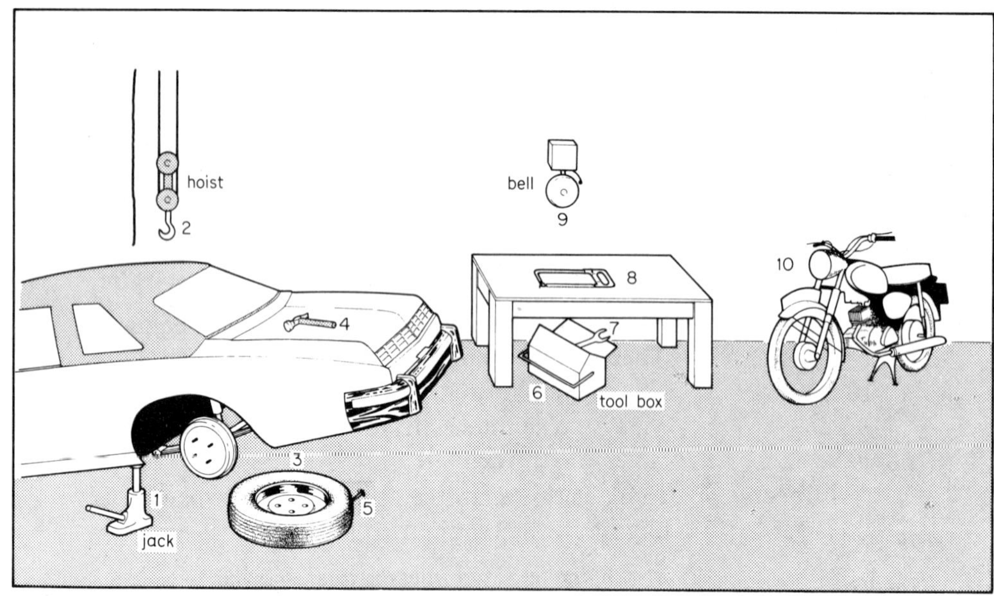

Now look at these examples.

There is a car in the garage.
There is a motorcycle in the garage.

Complete these sentences. Use *on*, *in*, *under*, *over* or *beside*.

1. There . . . . car.
2. There . . . . car.
3. There . . . . car.
4. There . . . . car.
5. There . . . . wheel.
6. There . . . . table.
7. There . . . . tool box.
8. There . . . . table.
9. There . . . . table.
10. There . . . . table.

Exercise 9 Make statements with the words below.

Example:

triangle / three sides	A triangle has three sides.
hammers / a blade	Hammers do not have a blade.

1. diesel engines / a fuel injector
2. petrol engines / spark plugs
3. 2-stroke petrol engines / an oil sump
4. a 4-stroke engine / valves
5. a 2-stroke petrol engine / valves
6. rectilinear shapes / angles
7. rectilinear shapes / curved sides
8. a hexagon / six sides
9. a rectangle / four equal sides
10. a parallelogram / right angles

Exercise 10 Look at the picture.

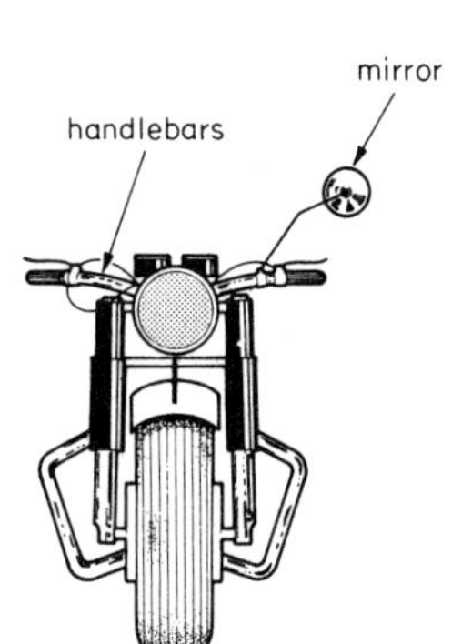

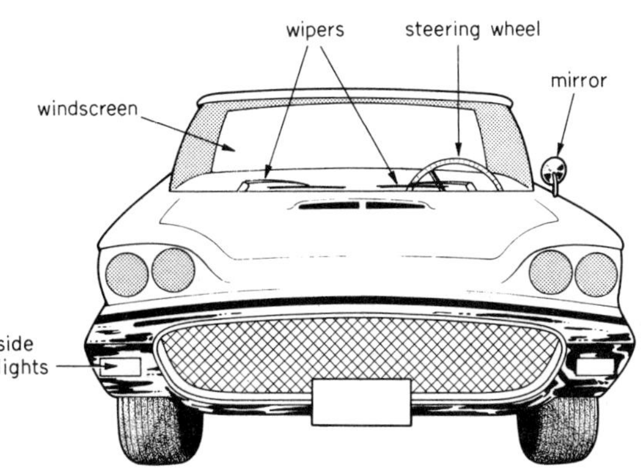

Complete this description.

In the diagram there is a motorcycle on . . . . and there is a . . . . on the right. . . . . a headlight on the front of the motorcycle. The car . . . . two headlights at . . . . There . . . . two wipers on the windscreen. . . . . no windscreen on the motorcycle and it has . . . . wipers. . . . . two side lights on the front of . . . . no side lights on the front of . . . .

. . . . a steering wheel in the car. A motorcycle . . . . no steering wheel. . . . . handlebars.

. . . . a mirror on the car and . . . . a mirror on the motorcycle.

LANGUAGE NOTE 15
at the front AND in the middle AND on the left

LANGUAGE NOTE 16
the spare wheel OR the car wheel BUT all the wheels both the wheels

a type	a jack	diesel
a stroke	a bell	fuel
a spark plug	a table	both
a fuel injector	a mirror	on the left
a valve	a wiper	on the right
a sump	a windscreen	under
a port	a steering wheel	over
a hoist	handlebars	beside

SECTION C: ENGINES AND FUELS

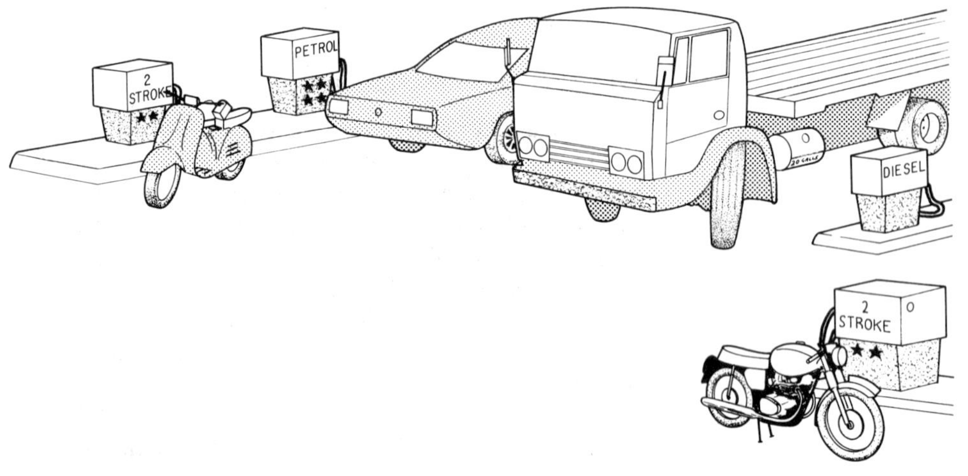

There are three basic types of fuel for motor vehicles. There is diesel fuel, there is petrol and there is two-stroke mixture.

Most motorcycles have two-stroke petrol engines. There are a few motorcycles with four-stroke petrol engines. There are no motorcycles with diesel engines.

A few cars have diesel engines. There are also a few cars with two-stroke engines but most cars have four-stroke petrol engines. Most lorries have diesel engines. There are a few lorries with four-stroke petrol engines but there are no lorries with two-stroke petrol engines.

There is a lot of carbon (about 85%) in diesel fuel. There is also a lot of carbon in petrol. There is a little hydrogen in both these fuels. There is also a little sulphur (about 1%) in diesel fuel but there is no sulphur in petrol.

Two-stroke mixture is about 95% petrol and 5% oil. There is no oil in the fuel for four-stroke engines.

Exercise 11 Answer these questions and put a tick (√) in the correct box. Here is an example.

How many motorcycles have diesel engines?
Short answer: None.
Complete answer: No motorcycles have diesel engines.

MOST	A FEW	NONE	A LITTLE	A LOT
		√		

1. How many motorcycles have four-stroke petrol engines?
2. How many motorcycles have two-stroke engines?
3. How many cars have diesel engines?
4. How many cars have two-stroke engines?
5. How many lorries have diesel engines?
6. How much carbon is there in diesel fuel?
7. How much hydrogen is there in petrol?
8. How much sulphur is there in diesel fuel?
9. How much oil is there in two-stroke mixture?
10. How much oil is there in petrol?

	MOST	A FEW	NONE	A LITTLE	A LOT
1.					
2.					
3.					

continued on page 46

	MOST	A FEW	NONE	A LITTLE	A LOT
4.					
5.					
6.					
7.					
8.					
9.					
10.					

Exercise 12 Make statements with these words.

Example: motorcycles / four-stroke petrol engines
There are a few motorcycles with four-stroke petrol engines.

1. cars / two-stroke engines
2. lorries / two-stroke engines
3. oil / diesel fuel
4. carbon / petrol
5. hydrogen / diesel fuel
6. sulphur / petrol
7. hammers / plastic heads
8. cars / three wheels
9. lorries / six wheels
10. motorcycles / air-cooled engines

Exercise 13 Look at these examples.

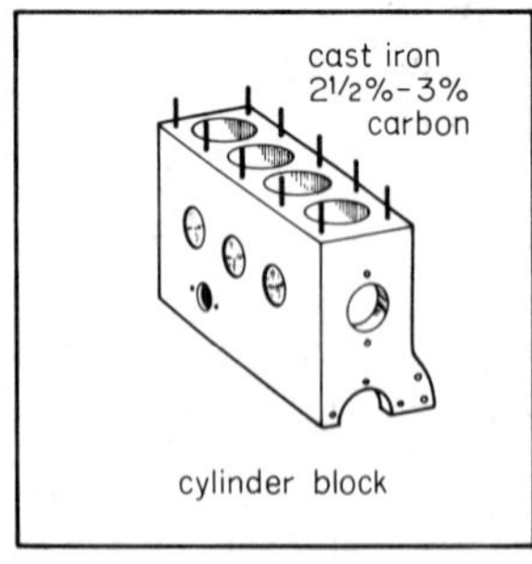

This cylinder block is made of cast iron.
There is a little carbon in cast iron.

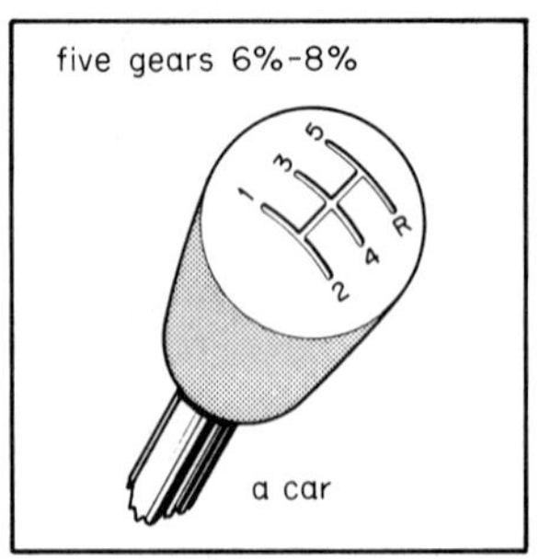

This car has five gears.
There are a few cars with five gears.

Now make two sentences about each picture in the same way.

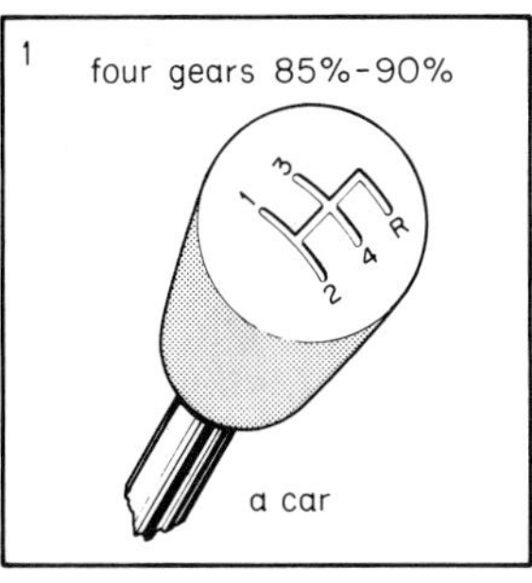

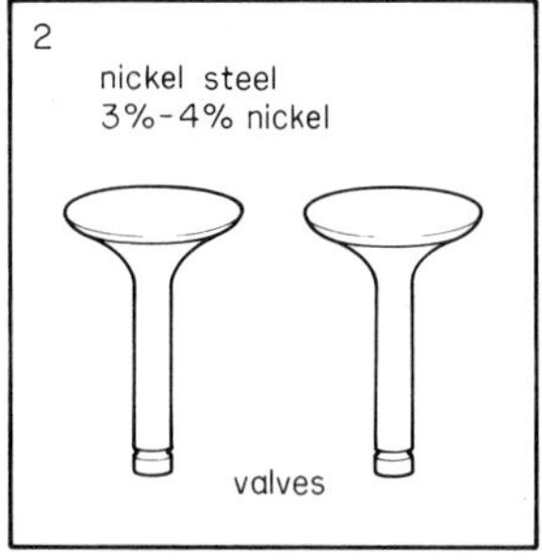

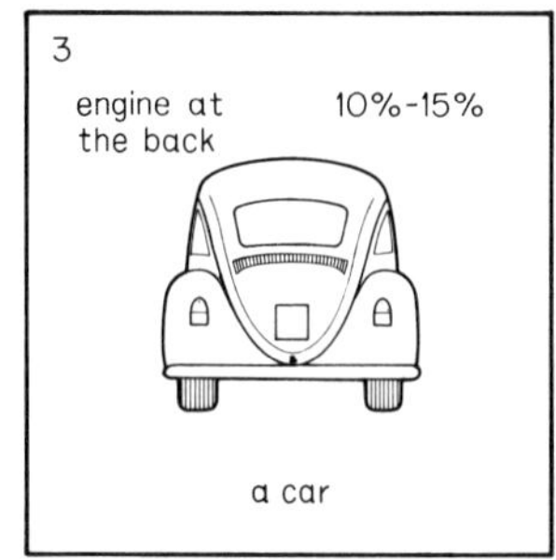

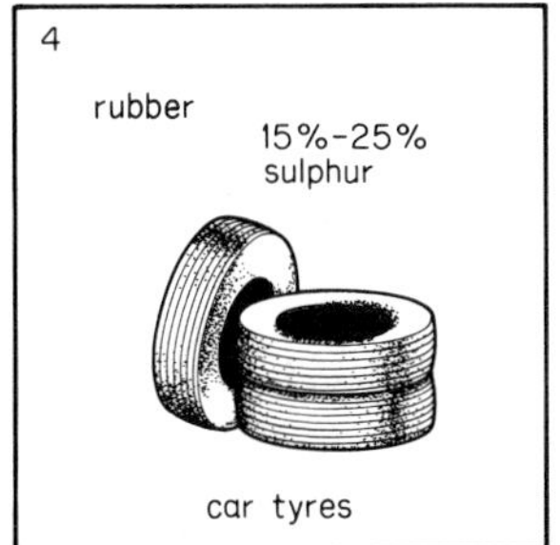

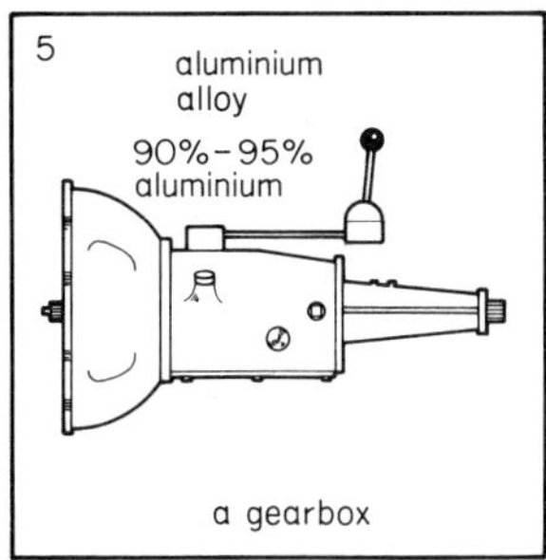

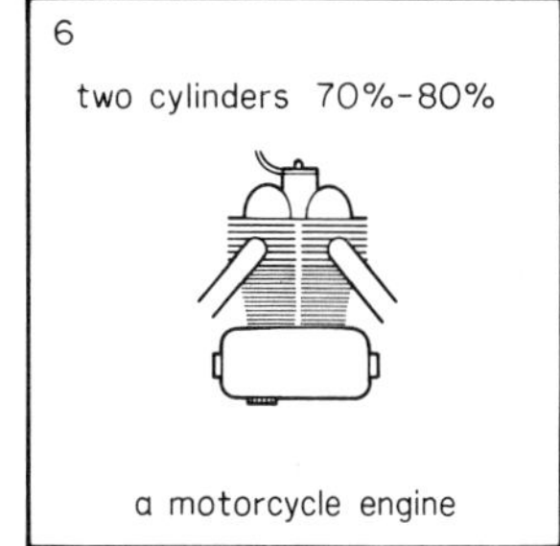

Exercise 14 Look at this diagram of an air-cooled 2-stroke petrol engine.

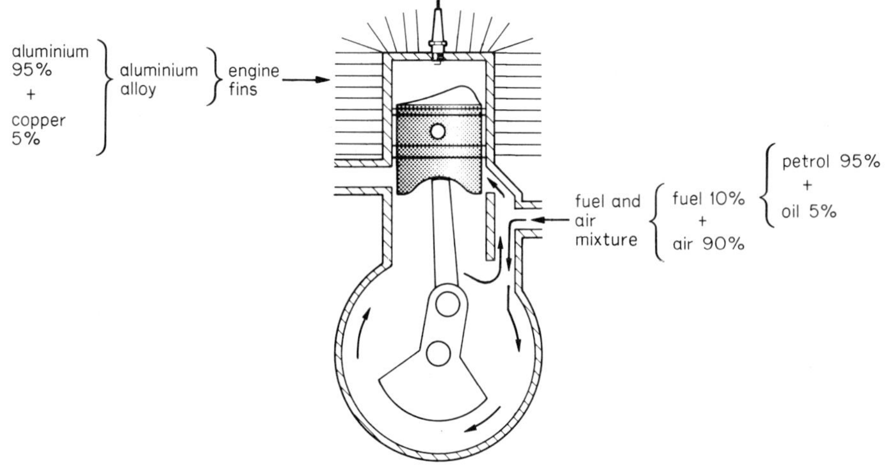

Now complete this description.

There . . . . a 2-stroke petrol engine . . . . the diagram. The engine fins . . . . made . . . . aluminium alloy. Aluminium alloy . . . . made . . . . two metals. There is . . . . aluminium (about 95%) and . . . . copper in most aluminium alloys. . . . . motorcycle engines (about 90%) have aluminium alloy fins.
There . . . . oil sump in this engine. There is . . . . oil in the petrol. There is . . . . air and . . . . fuel in the fuel and air mixture.

Exercise 15 *Amounts* are often expressed as *percentages* or *fractions.*

Complete this table. Some of them have been done for you.

PERCENTAGES	FRACTIONS
50% (fifty per cent)	$\frac{1}{2}$ (a half)
25% (twenty five per cent)	$\frac{1}{4}$ (a quarter)
75% ( )	(three quarters)
(thirty three and a third per cent)	$\frac{1}{3}$ ( )
(twenty per cent)	(a fifth)
(forty per cent)	( )
(seventeen per cent)	( )
10% ( )	( )
(sixteen and two thirds per cent)	$\frac{1}{6}$ ( )
(two per cent)	(a fiftieth)
1% ( )	( )

LANGUAGE NOTE 17		
2nd = second	BUT	$\frac{1}{2}$ = a half
4th = fourth		$\frac{1}{4}$ = a quarter

LANGUAGE NOTE 18	
five → fifth, fifteen, fifty	
lorry → lorries	
fifty → fiftieth	

a mixture
a lorry
a block
a gear
a gearbox
an alloy
a tyre
a percentage
a fraction

carbon
hydrogen
sulphur
cast iron
nickel
steel
aluminium
copper
rubber
air

basic
most
a few
a little
a lot (of)
none

per cent (%)

with
about

how much . . . . ?

UNIT FOUR Instructions

SECTION A: A FIRE IN THE WORKSHOP

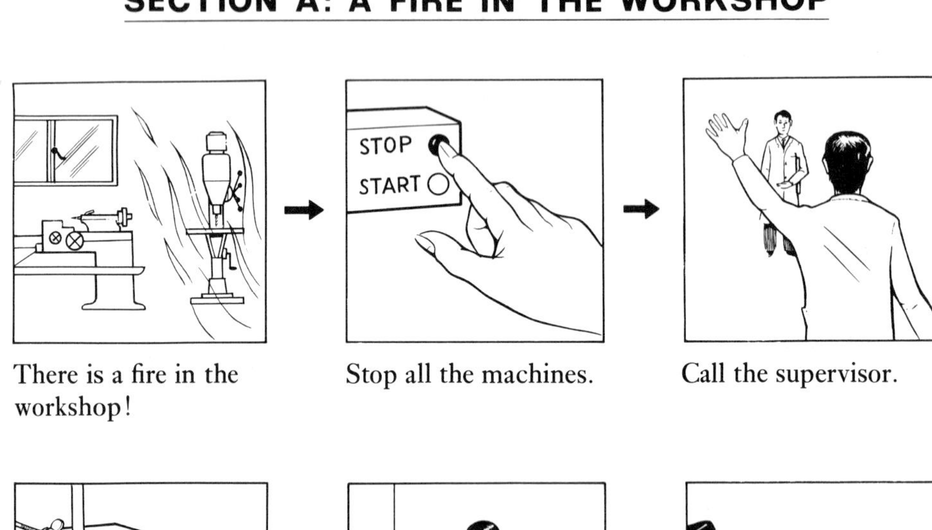

There is a fire in the workshop!

Stop all the machines.

Call the supervisor.

Turn off the mains switch.

Take a fire extinguisher from the wall.

Read the instructions on the extinguisher.

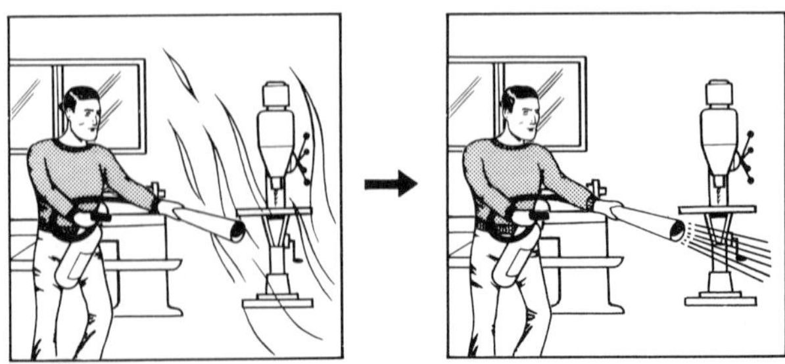

Point the extinguisher at the fire.

Extinguish the fire.

Exercise 1 Complete these instructions with the correct verb.

A. . . . . a fire extinguisher from the wall.
B. . . . . the instructions on the fire extinguisher.
C. . . . . the machines in the workshop.
D. . . . . the mains switch.
E. . . . . the supervisor.
F. . . . . the fire.
G. . . . . the extinguisher at the fire.

Now rewrite the instructions in the correct order.

Exercise 2 Complete these sentences.

1. There is a fire . . . . workshop.
2. There are six machines . . . . the workshop.
3. The mains switch is . . . . the wall.
4. There is a handle . . . . the mainswitch.
5. Point the fire extinguisher . . . . the fire.
6. The fire extinguishers are . . . . the wall.
7. Take a fire extinguisher . . . . the wall.
8. There are instructions . . . . the fire extinguisher.

Exercise 3 Here is a plan of a workshop.

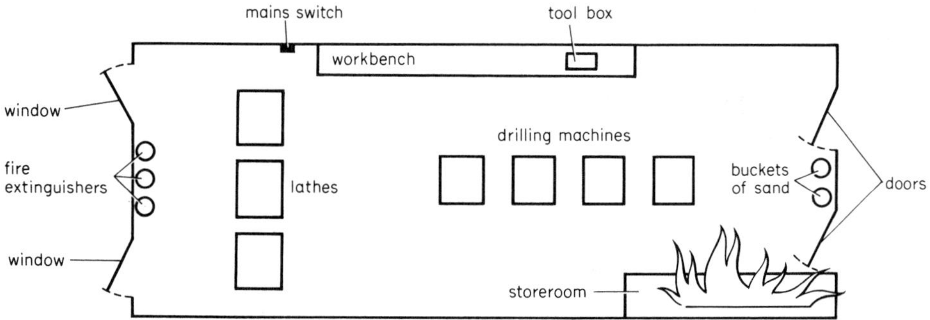

Look at this example.

Where are the buckets of sand?
They are between the doors.

Make six questions from this table. Make answers from the picture. Use words from the list.

between	*in*	*in the middle*
beside	*on*	*on the left*

Where	is are	the lathes? the drilling machines? the tool box? the mains switch? the fire extinguishers? the fire?

Exercise 4 Follow these instructions.

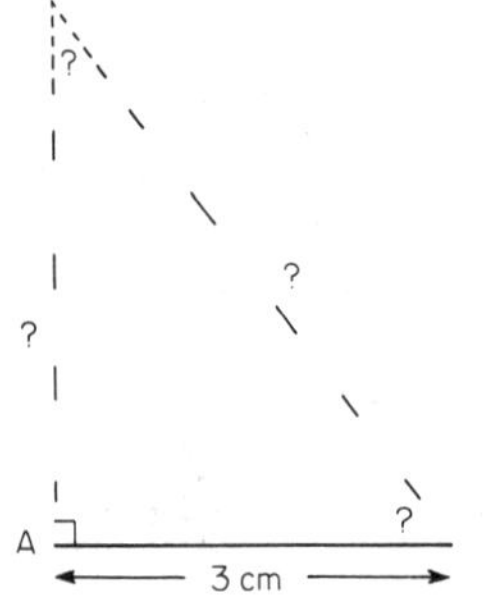

Complete this triangle.

1. Take a ruler, a pencil and a protractor.
2. Draw a side 3 cm long.
3. Label the side AB.
4. Make a right angle at A with the protractor.
5. Draw a side 4 cm long from A. Label the side AC.
6. Join B and C.

Now answer these questions:

1. How wide is angle C? (Measure the angle.)
2. How wide is angle B?
3. How long is the line CB?
4. How long is the line AC?

Exercise 5 Complete these instructions. Use these words.

draw	*diameter*
take	*join*
another	*protractor*
measure	*label*

1. . . . . . a circle.
2. . . . . . the centre X.
3. . . . . . a ruler and draw a . . . . AB.
4. Make a right angle at X with a . . . . .
5. Draw . . . . diameter CD.
6. . . . . . AC, CB, BD and AD.
7. . . . . . angle A. Is it a right angle?

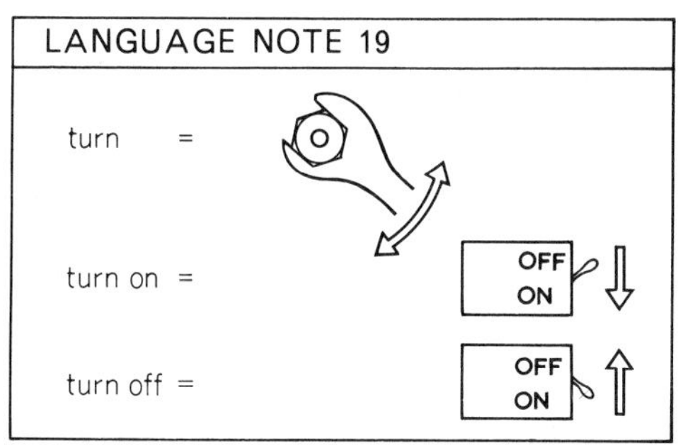

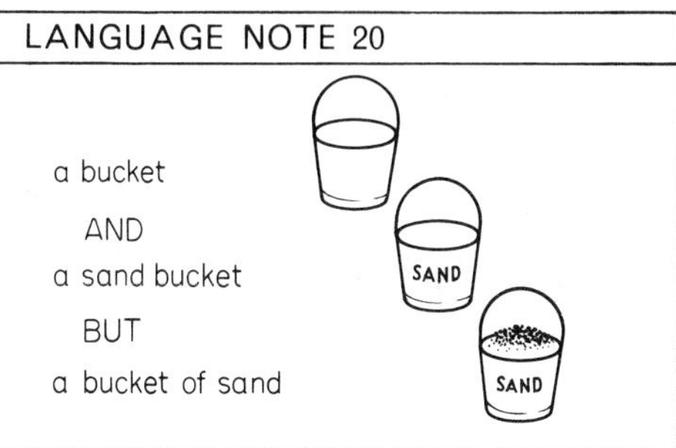

an instruction
a fire
a workshop
a machine
a supervisor
a mains switch
a fire extinguisher
a wall
a door
a window
a machine
a lathe
a workbench
a storeroom
a diameter

turn off
take
read
point
extinguish
make
join

at
from
between

where . . ?

SECTION B: DIFFERENT TYPES OF FIRES

Type A
Type B
Type C
Wood, paper, leather etc.
are flammable materials
Oil, petrol etc.
are flammable liquids
Electrical Equipment

Here are three types of fire. There are different instructions for each type.

Type A: Flammable Materials

1. First, remove all dangerous materials near the fire.
2. Next, close all windows and ventilators.
3. Then, throw water over the fire.

Type B: Flammable Liquids

1. First, remove all containers to a safe place.
2. Next, close all windows and ventilators.
3. Then, use the correct fire extinguisher. Do not throw water over this type of fire. Extinguish it with CO_2 or foam.

Type C: Electrical Equipment

1. First, switch off the equipment at the mains.
2. Then, use the correct fire extinguisher. Do not use water on an electrical fire. Do not use foam on it. Extinguish it with CO_2.

Exercise 6 Some of these instructions are incorrect. Put '*Do not . . .*' in the incorrect sentences.

Flammable Materials

1. Close all windows in the room.
2. Use water on flammable materials.

Flammable Liquids

3. Remove all oil, petrol containers etc to a safe place.
4. Use water on flammable liquids.
5. Use foam on flammable liquids.

Electrical Equipment

6. Use water on electrical fires.
7. Use foam on electrical fires.
8. Use CO_2 on electrical fires.

Exercise 7 Look at this example.

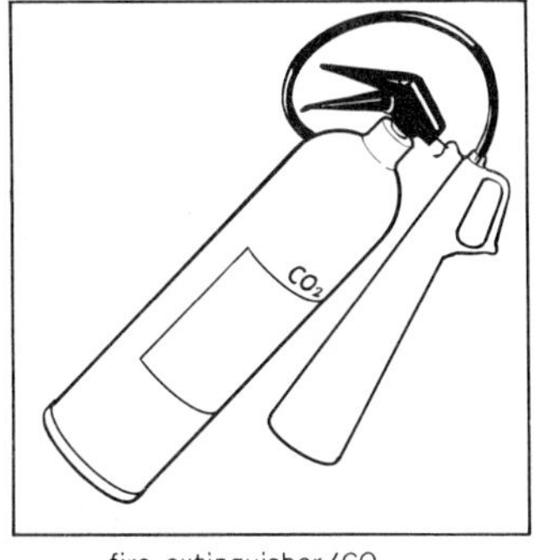

fire extinguisher/CO_2

This fire extinguisher has CO_2 in it.

Now make sentences from these pictures in the same way.

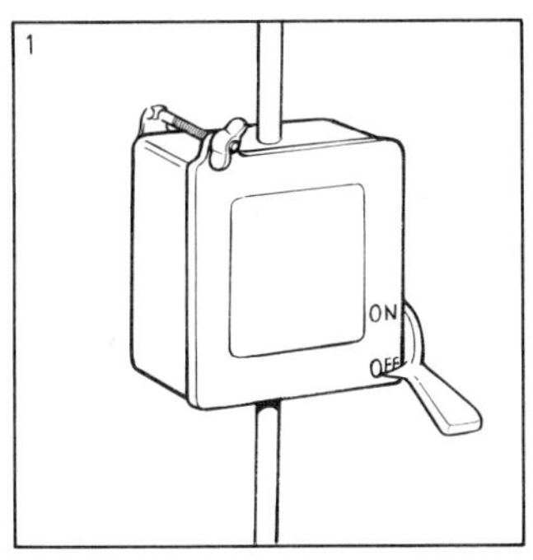

mains switch/handle

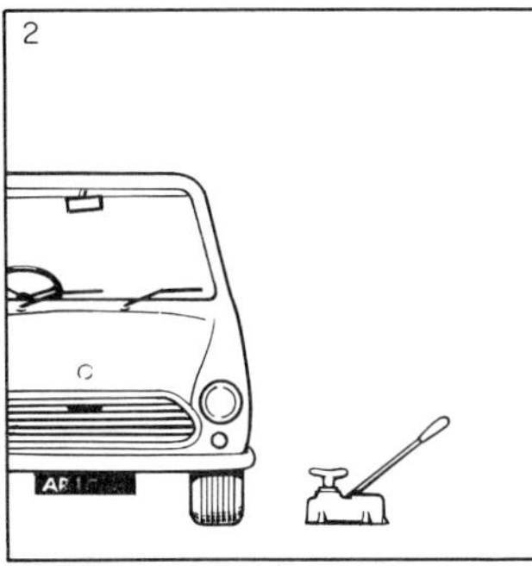

car/jack

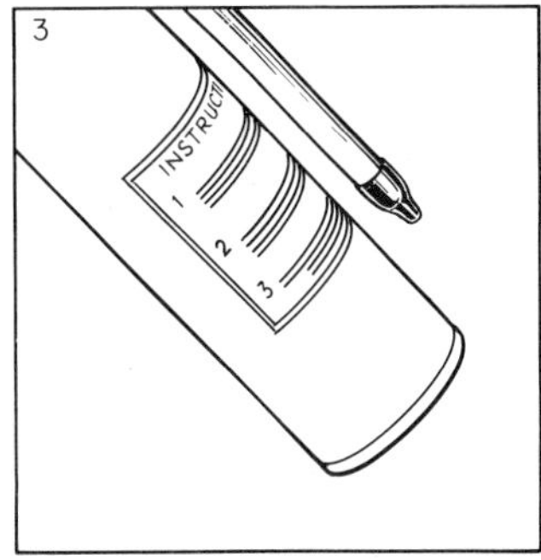

fire extinguisher/instructions

bucket/sand

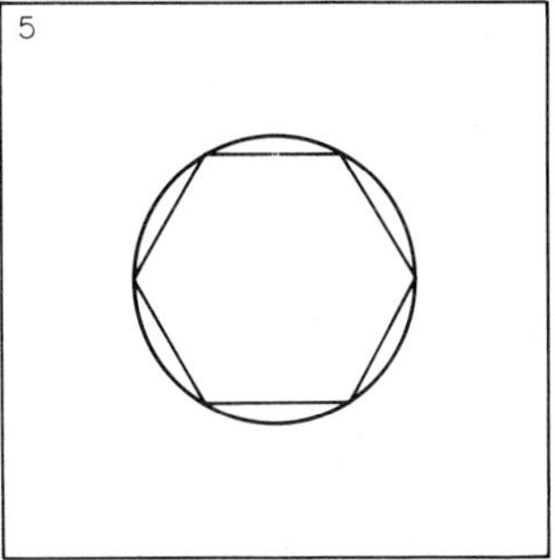

circle/hexagon

workbench/wheel

Exercise 8 Write eight instructions from the pictures for extinguishing an electrical fire.

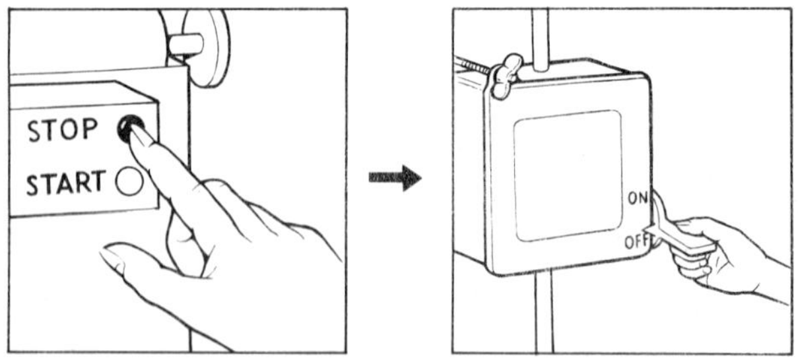

Exercise 9 Follow these instructions.

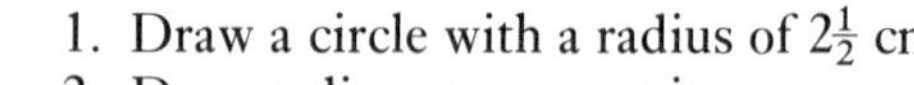
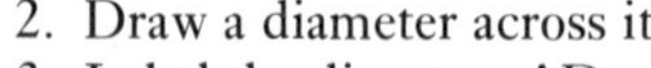

1. Draw a circle with a radius of $2\frac{1}{2}$ cm.
2. Draw a diameter across it.
3. Label the diameter AD.
4. Place the compass point on A.
5. Draw an arc across the circle with the same radius ($2\frac{1}{2}$ cm.)
6. Label the arc BF.
7. Draw another arc with the compass point on D. Do not change the radius.
8. Label this arc CE.
9. Join AB, BC, CD, DE, EF and FA.

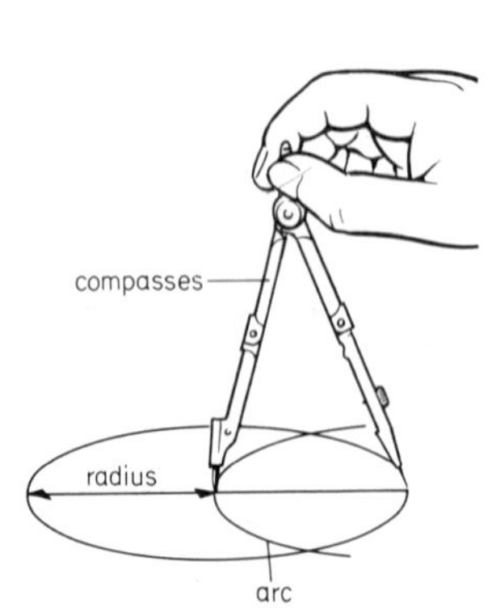

Now answer these questions.

How many angles has this shape?
What shape is it?
How wide is angle A?
Are all the angles equal?
How long is the side AB?
Do all the sides have the same length?

Exercise 10 Complete these instructions for changing a wheel from the words in this list.

from	*replace*	*to*
nuts	*spanner*	*under*
on	*take*	*wheel*
remove	*tighten*	*with*

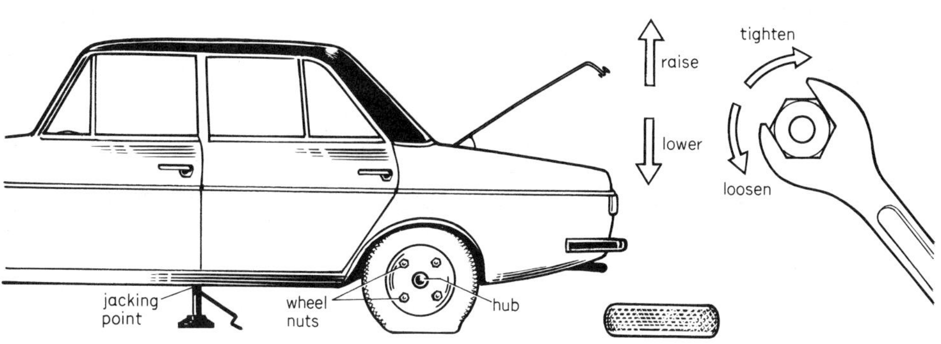

— Take the tools and the spare wheel . . . . the car.
— Place the jack . . . . the jacking point.
— Loosen the wheel nuts with a . . . .
— Raise the wheel . . . . the jack.
— . . . . the wheel nuts.
— . . . . the wheel from the hub.
— Place the spare wheel . . . . the hub.
— Replace the wheel . . . .
— Lower the wheel . . . . the ground.
— . . . . the nuts.
— . . . . the tools in the car.

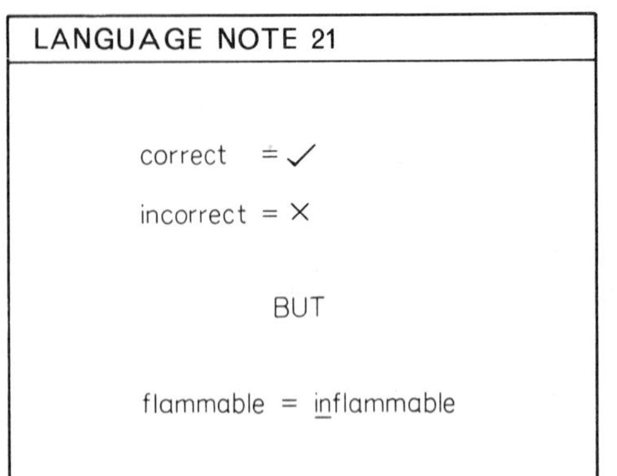

a ventilator
a place
a liquid
a radius
a point
an arc
a hub

(the) ground
equipment
CO_2 (carbon dioxide)
foam
compass(es)

remove
close
throw
switch off
use
change
(re)place
loosen
tighten
raise
lower

at the mains

different
each
dangerous
safe
flammable
correct
electrical

near
to
across

then
etc. (etcetera)

SECTION C: FIRE EXTINGUISHERS

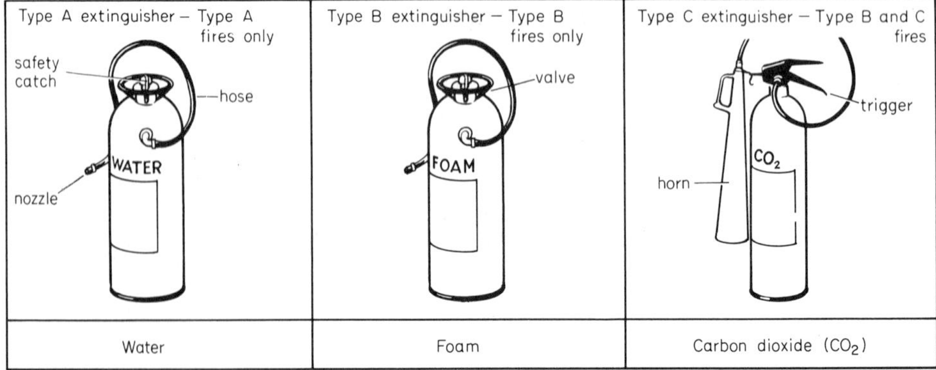

Here are three main types of fire extinguisher. They have different materials in them.
Note: Always read the instructions on the extinguisher.

Type A: These extinguishers are always red. They have water in them. They have a rubber hose. Use these extinguishers on Type A fires only.

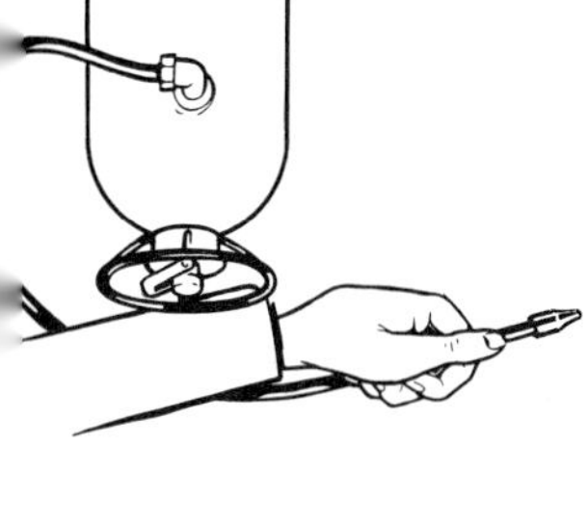

Instructions

1. Hold the nozzle in one hand.
2. Point the nozzle at the fire.
3. Remove the pin.

Type B: These extinguishers have foam in them. They are often white. They sometimes have a nozzle and they sometimes have a horn. They always have a valve.

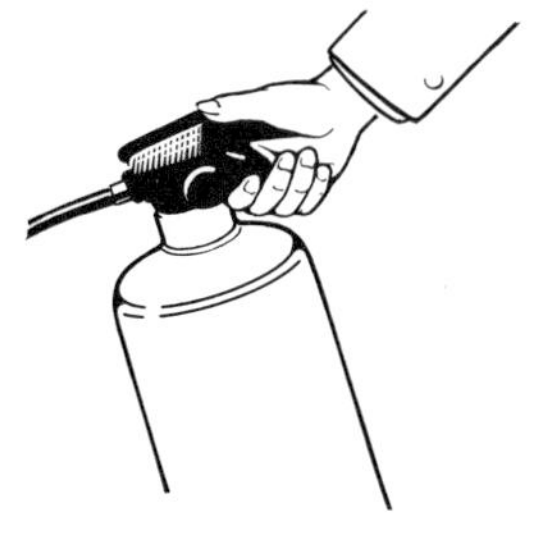

Instructions

1. Hold the horn or nozzle in one hand.
2. Open the valve.
3. Spray the foam over and around the fire.

Note: Never use foam extinguishers on electrical fires.

Type C: These extinguishers have CO_2 in them. They are usually black. They have a large horn. *Never* use these extinguishers without a horn.

Instructions

1. Point the horn at the fire.
2. Press the trigger.

Note: Do *not* use CO_2 extinguishers on Type A fires.

Exercise 11 Are these statements true or false?

1. Type A extinguishers have water in them.
2. Type A extinguishers are never red.
3. Type B extinguishers are usually black.
4. There is foam in Type B extinguishers.
5. Type C extinguishers are always white.
6. Foam extinguishers never have a horn.
7. Type A extinguishers always have a horn.
8. Use Type C extinguishers on electrical fires.

Exercise 12 Put *always*, *often*, *usually*, *only*, *never* or *sometimes* in these sentences.

1. Use CO_2 extinguishers without a horn.
2. Read the instructions on the extinguisher.
3. Foam extinguishers are white.
4. Use water on Type A fires.
5. Type A extinguishers are red.
6. Foam extinguishers have a nozzle.
7. Type C extinguishers are black.
8. Use CO_2 extinguishers on Type A fires.

Exercise 13 Make an instruction from each of these pictures. Here are two examples.

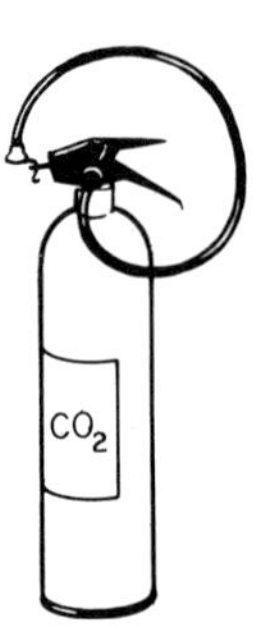

Never use a CO_2 extinguisher *without* a horn.

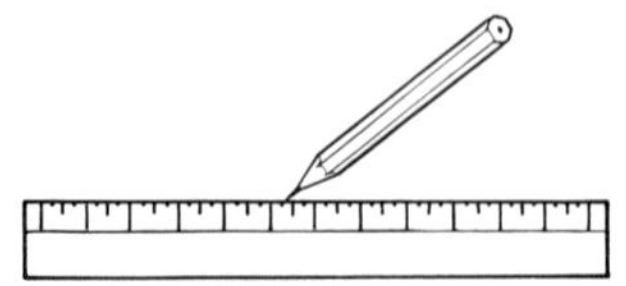

Always use a ruler *for* draw*ing* straight lines.

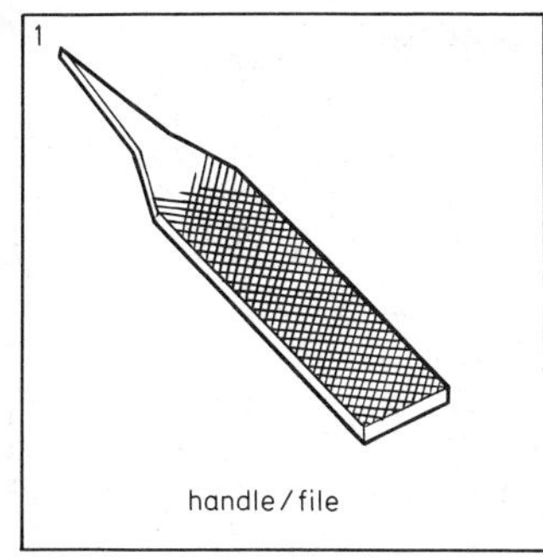
1
handle / file

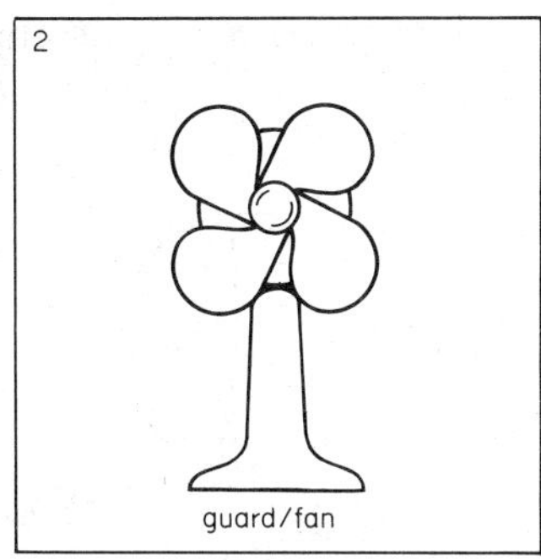
2
guard/fan

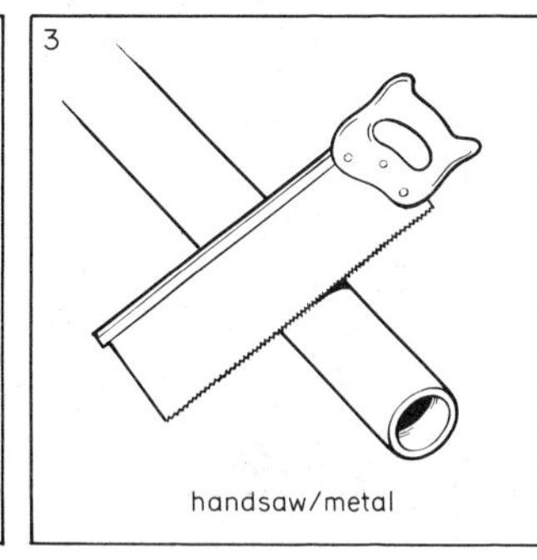
3
handsaw/metal

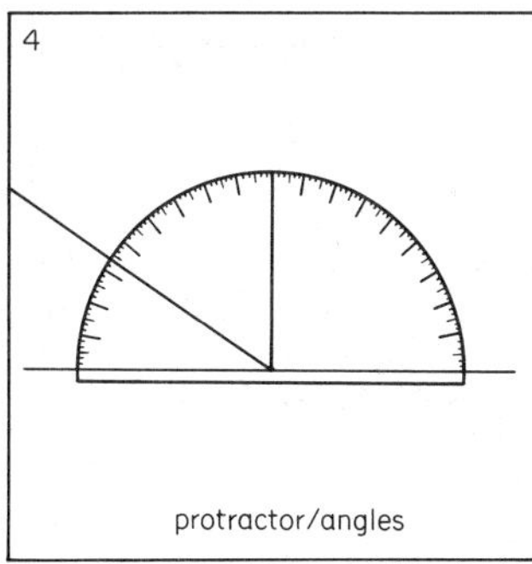
4
protractor/angles

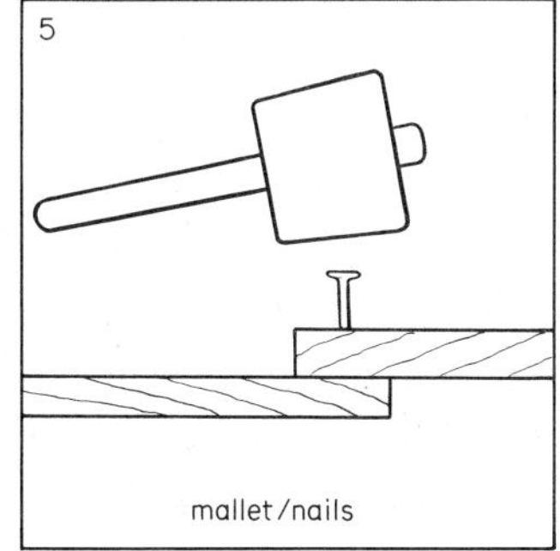
5
mallet/nails

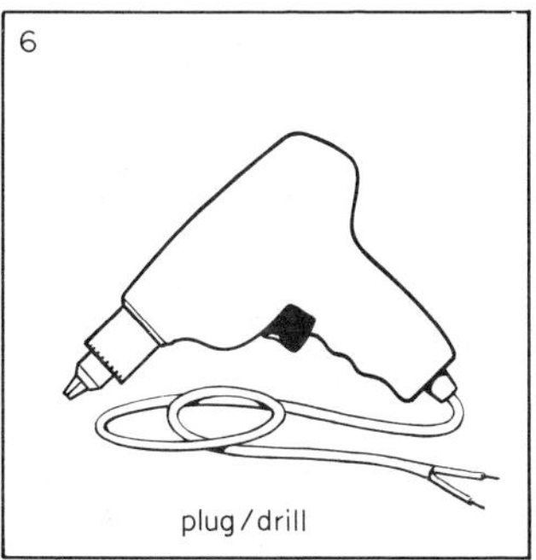
6
plug/drill

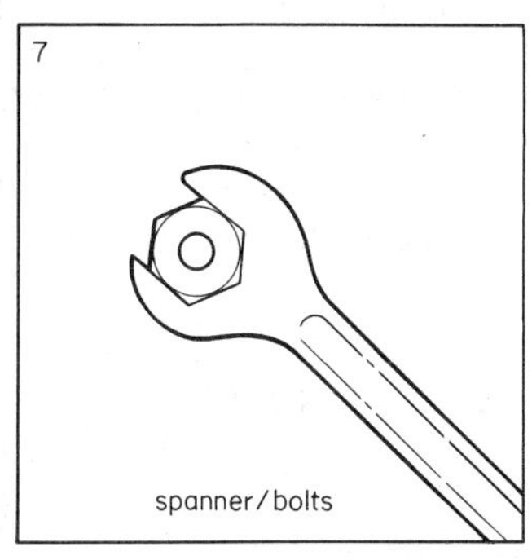
7
spanner/bolts

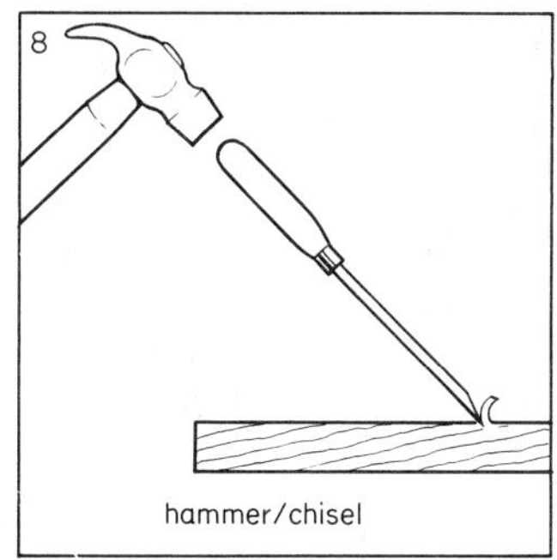
8
hammer/chisel

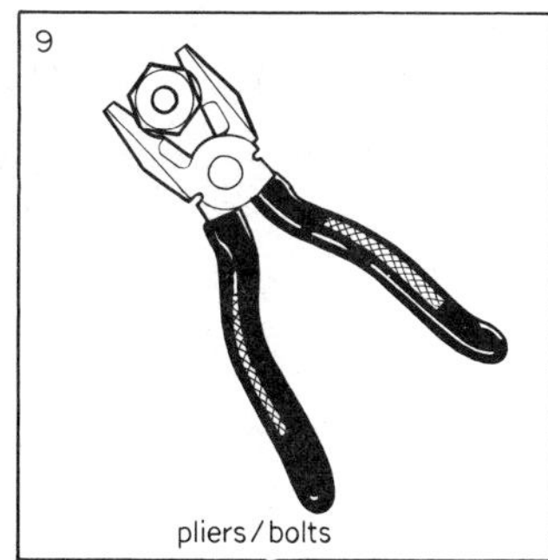
9
pliers/bolts

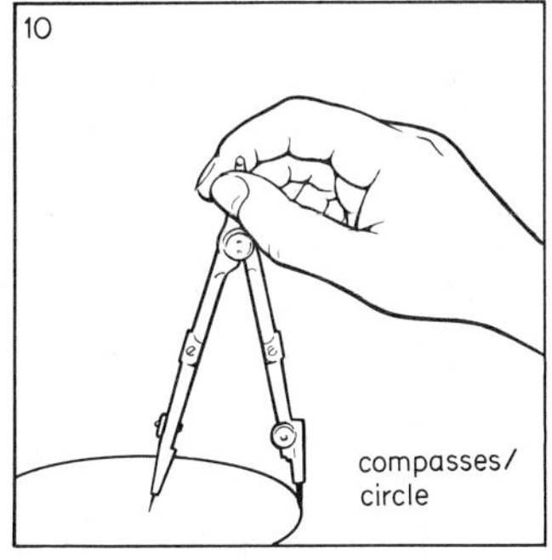
10
compasses/
circle

Exercise 14 Look at the picture and the example under it.

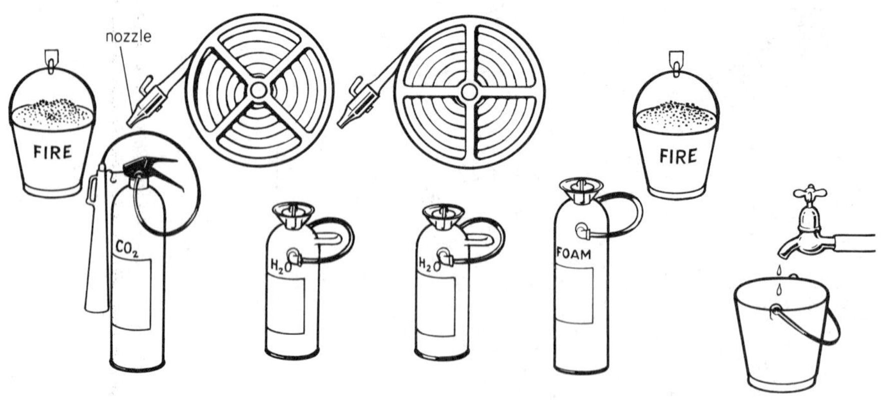

There are three types of extinguisher in the workshop.
They have different materials in them.

Complete these sentences in the same way.

1. There are two red fire extinguishers in the middle.
   They have . . . .

2. There is a black extinguisher on the left.
   It . . . .

3. There is a white extinguisher on the right.
   . . . .

4. There are two hoses over the extinguishers.
   . . . .

5. There is a tap on the right.
   . . . . under . . . .

6. There are two buckets on the wall.
   . . . .

Exercise 15 What are the three instructions on this Type B extinguisher?

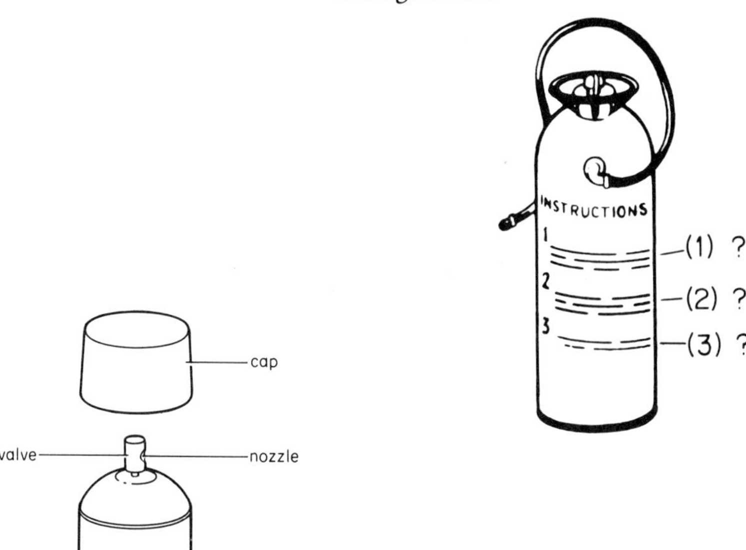

Here is a can of fly spray.

Make six instructions from the pictures.

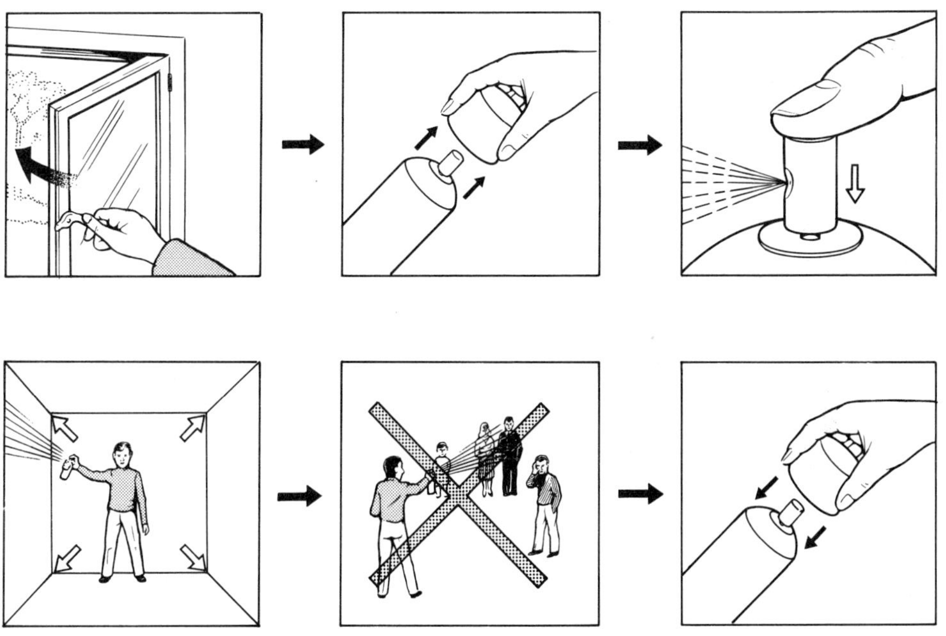

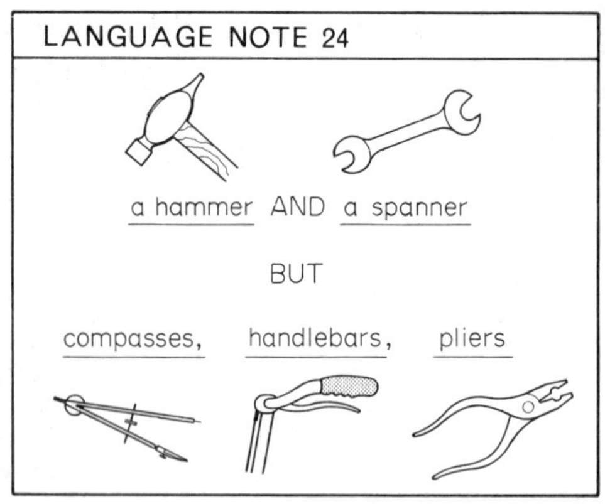

a hose
a nozzle
a pin
a hand
a trigger
a horn
a guard
a plug

pliers

open
press
spray

note

without

main
red
white
black
large

only
often
usually

UNIT FIVE Facts About Matter

SECTION A: THREE STATES OF MATTER

Under normal conditions, there are three states of matter: the solid state, the liquid state and the gas state. Iron, nickel and other metals are normally solids. Water is normally a liquid and carbon dioxide is normally a gas.

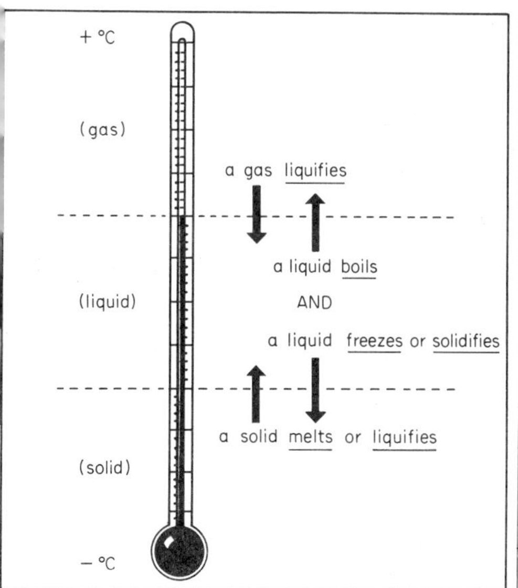

fig. (a)

METALS	m.p.
Iron (Fe)	1,535°C
Nickel (Ni)	1,453°C
Tin (Sn	232°C

fig. (b)

LIQUIDS	f.p.	b.p.
Water (H_2O	0°C	100°C
Bromine (Br)	−7°C	58°C
Benzene (C_6H_6)	5°C	80°C

fig. (c)

GASES	f.p.	b.p.
Carbon Dioxide (CO_2)	−78°C	−57°C
Oxygen (O)	−219°C	−183°C
Hydrogen (H)	−259°C	−253°C

fig. (d)

Fig. (b) gives the melting points (m.p.) of some metals. Iron has a high melting point. It melts at 1,535°C. Nickel has a high m.p. Tin has a low m.p. It melts at 232°C.

Fig. (c) gives the freezing points (f.p.) and the boiling points (b.p.) of liquids. Water is normally a liquid. It boils at 100°C, *i.e.* it becomes a gas above this temperature. It freezes at 0°C, *i.e.* it becomes a solid below this temperature. Bromine is normally a liquid but it has a low b.p. It boils at 58°C and it freezes at −7°C.

Fig. (d) gives the f.p. and the b.p. of gases. At low temperatures, a gas becomes a liquid. Oxygen liquifies at −183°C. Carbon dioxide is normally a gas but it is often used in the liquid or the solid state. CO_2 liquifies at −57°C. It solidifies at −78°C.

Exercise 1 Are these statements true or false?

1. The solid state, the liquid state and the gas state are the three normal states of matter.
2. Metals are normally liquids.
3. Iron liquifies at 1,535°C.
4. Tin has a high m.p.
5. Water boils at 0°C and it freezes at 100°C.
6. Bromine becomes a gas *above* −7°C.
7. Benzene becomes a gas *above* 80°C.
8. CO_2 normally solidifies *below* −78°C.
9. Oxygen normally becomes a liquid *above* −183°C.
10. Hydrogen is in the liquid state between −253°C and −259°C.

Exercise 2 Complete these sentences.

1. Fig. (b) . . . . the m.p. of some metals.
2. Fig (c) . . . . the f.p. and . . . .
3. Fig. (d) . . . .
4. Iron . . . . at 1,535°C.
5. Nickel . . . . *i.e.* it . . . . a liquid above this temperature.
6. Tin . . . . C, *i.e.* it . . . .
7. Water . . . . at 0°C, *i.e.* it . . . .
8. Benzene . . . . 80°C, *i.e.* . . . .
9. Oxygen . . . . at −183°C. and it . . . . at −219°C.
10. Hydrogen . . . . at −253°C, *i.e.* . . . . and it . . . . at . . . . °C, *i.e.* . . . .

Exercise 3 Look at these tables and the example.

METALS	m.p.
Chromium (Cr)	1,900°C
Copper (Cu)	1,083°C
Aluminium (Al)	660°C
Lead (Pb)	327°C
Mercury (Hg)	−39°C

fig. (e)

LIQUIDS	b.p.
Sulphuric acid (H_2SO_4)	338°C
Nitric acid (HNO_3)	83°C
Carbon tetrachloride (CCl_4)	77°C

fig. (f)

GASES	b.p.
Carbon monoxide (CO)	−191°C
Sulphur dioxide (SO_2)	−10°C
Chlorine (Cl)	−34°C

fig. (g)

At normal temperatures, chromium is a *solid*. It *melts* at 1,900°C, *i.e.* it becomes a *liquid above* this temperature.

Now make ten short paragraphs from the tables in the same way.

Exercise 4 Look at this example.

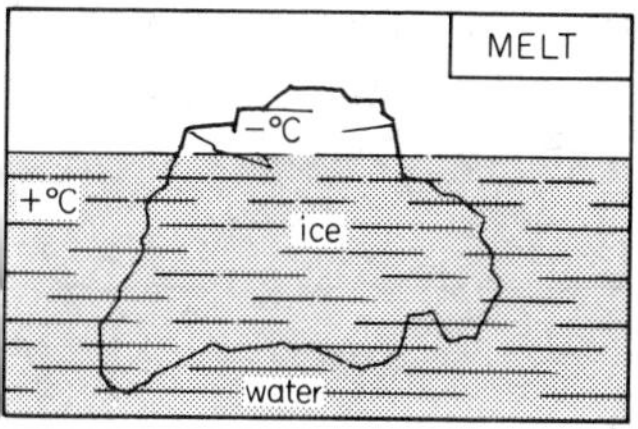

Water melts ice.

Now make sentences in the same way from these pictures.

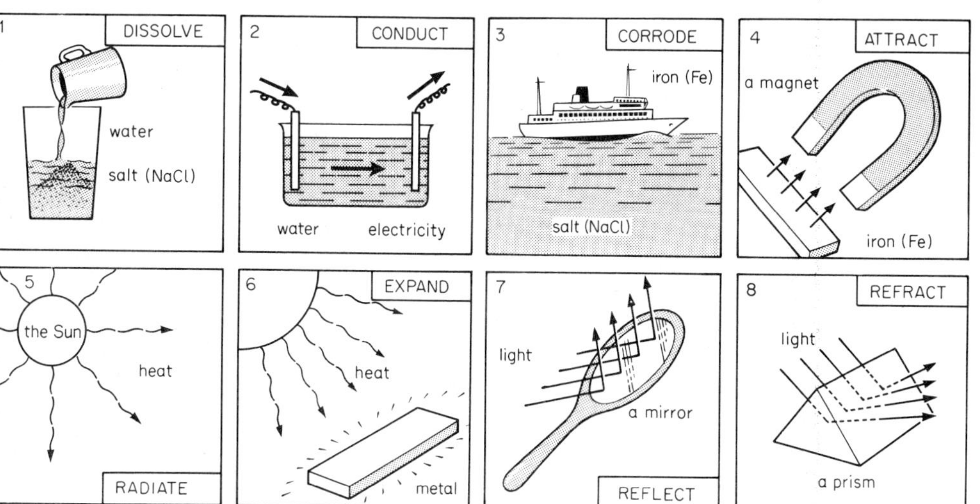

Exercise 5 Complete the following two paragraphs from the wordlist.

	m.p.	b.p.
Mercury (Hg)	−39°C.	357°C.

solid	*conditions*	*above*
liquid	*melts*	*below*
gas	*boils*	*between*
high	*freezes*	*measuring*
low	*becomes*	*used*

Mercury is a metal. Under normal . . . . , it is also a liquid. It has a . . . . m.p. It . . . . at −39°C, *i.e.* it is a . . . . below this temperature and it is a . . . . above this temperature. Liquid mercury is . . . . in thermometers.

A metal also has a b.p. At . . . . temperatures, a liquid metal . . . . a gas. Liquid mercury . . . . at 357°C, *i.e.* it is a liquid . . . . this temperature and it becomes a gas . . . . this temperature. Mercury thermometers are not used for . . . . temperatures above 350°C.

LANGUAGE NOTE 25

metal ✓		a metal ✓
gas ✓	AND	a gas ✓
liquid ✓		a liquid ✓
water ✓		a water ×
benzene ✓	BUT	a benzene ×
hydrogen ✓		a hydrogen ×
wood ✓		a wood ?
glass ✓	OR	a glass ?
iron ✓		an iron ?
nickel ✓		a nickel ?

LANGUAGE NOTE 26

a condition
a fact
a state
a figure (fig.)

a melting point (m.p.)
a freezing point (f.p.)
a boiling point (b.p.)

a magnet

the Sun

normal(ly)
low

above
below
about

Celsius (Centigrade)
zero (0)
minus (—)

ice
electricity
heat
light
matter

some

give
melt
boil
freeze
become
liquify
solidify
dissolve
conduct
corrode
radiate
reflect
expand
refract

i.e.

SECTION B: THE BOILING POINT OF WATER

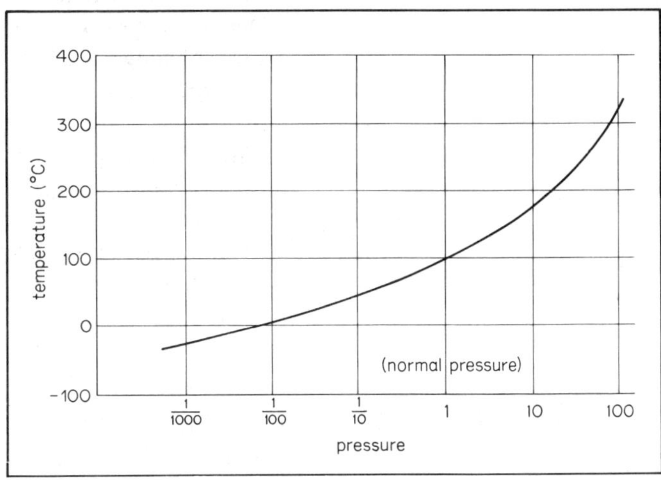

fig. (**h**) the b.p. of Water (approximate)

Fig. (h) is a graph. It shows the boiling point of water at different pressures. The vertical scale gives different temperatures and the horizontal scale gives different pressures. On the horizontal scale, one (1) is normal pressure.

At normal pressure, pure water boils at exactly 100°C. It does not always boil at exactly 100°C. The b.p. changes with pressure. The graph line shows the b.p. at different pressures.

At low pressures, water has a low b.p. For example, at normal pressure, water does not boil below 100°C, but at one tenth normal pressure, it boils at approx. 50°C. At high pressures, water has a high b.p. At normal pressure, it does not have a b.p. above 100°C, but at ten times normal pressure, the b.p. is approx. 185°C.

Note: The freezing point does not change with pressure. Pure water always freezes at exactly 0°C.

Exercise 6 Are these statements true or false? If they are false, change the statement into the negative form.

1. Pure water normally boils at exactly 100°C.
2. Water always boils at exactly 100°C.
3. Fig. (h) shows the b.p. of water at different pressures.
4. The graph line shows the b.p. of water at different pressures.
5. The horizontal scale gives different temperatures.
6. At normal pressure, water boils below 100°C.
7. At low pressures, water boils at exactly 100°C.
8. At high pressures, water has a b.p. of 100°C.
9. At a hundred times normal pressure, water boils at approx. 300°C.
10. The f.p. of pure water changes with pressure.

Exercise 7 Look at **fig. (i)** and complete the paragraphs from the wordlist. Complete the shaded boxes (▒) from the graph.

solid	*graph*	*horizontal*
liquid	*CO_2*	*vertical*
gas	*liquifies*	*normal*
temperature	*solidify*	*approximate*
pressures	*changes*	*times*

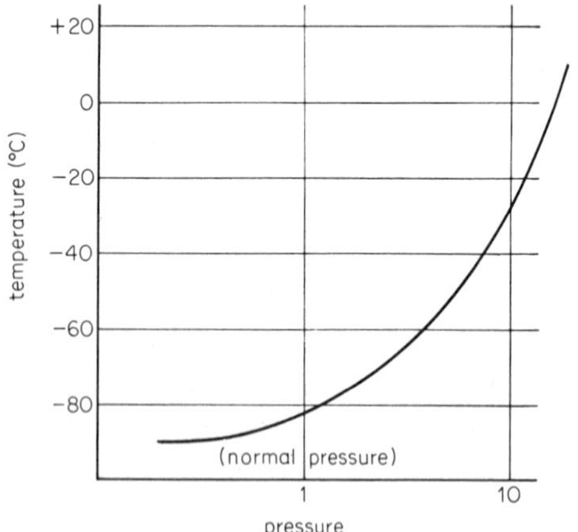

fig. (i) the b.p. of CO_2

Fig. (i) is a . . . . It shows the b.p. of . . . . at different pressures. The . . . . scale gives different temperatures in °C. The . . . . scale gives different pressures. On the horizontal scale, one (1) is . . . . pressure.

A gas does not always liquify at the same . . . . The b.p. . . . . with pressure. The graph line shows the b.p. at different . . . .

At normal pressure, CO_2 . . . . at approx. ▒ °C, *i.e.* it is a . . . . above this temperature and it becomes a . . . . below this temperature. At ten . . . . normal pressure, CO_2 liquifies at approx. ▒ °C.

Exercise 8 Look at these two examples.

a) Does the b.p. of water change with pressure?
Yes, it does.

b) Does the f.p. of water change with pressure?
No, it doesn't.

Now look at the diagram about water. Answer the questions.

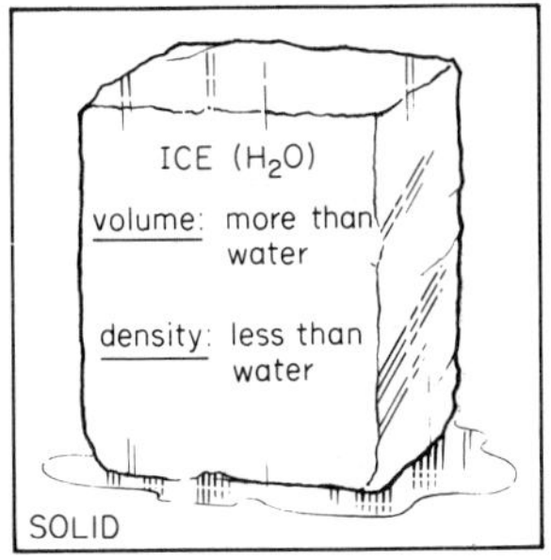

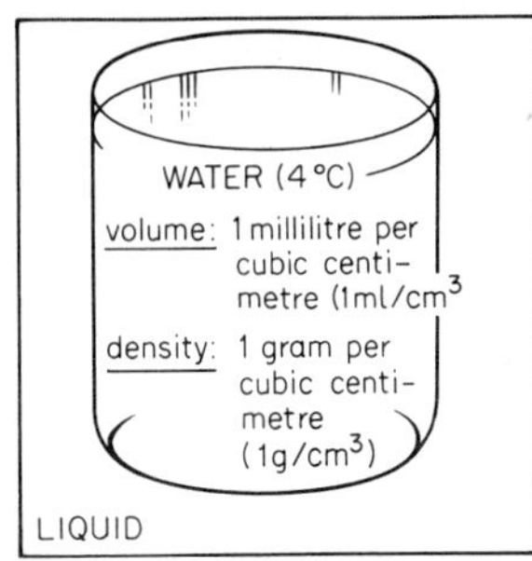

Does the	volume density	of water change with temperature?

Does	ice water steam	have	a volume a density	of	more than less than exactly	1 ml/cm^3 1 g/cm^3	?

Exercise 9

Exercise 10 Complete this table. Some of them have been done for you.

$\frac{1}{2}$ = a half			
= a third		$\frac{2}{3}$ =	
$\frac{1}{4}$ =		= three quarters	
$\frac{1}{5}$ =	= two fifths	$\frac{3}{5}$ =	= four fifths
= a sixth		$\frac{5}{6}$ =	
$\frac{1}{8}$ =	$\frac{3}{8}$ =	= five eighths	= seven eighths
$\frac{1}{10}$ =	$\frac{?}{10}$ =	$\frac{?}{10}$ =	$\frac{?}{10}$ =
$\frac{1}{16}$ =	$\frac{?}{16}$ =	$\frac{?}{16}$ =	$\frac{?}{16}$ =
$\frac{3}{16}$ =	$\frac{?}{16}$ =	$\frac{?}{16}$ =	$\frac{?}{16}$ =
$\frac{1}{100}$ =		= a thousandth	

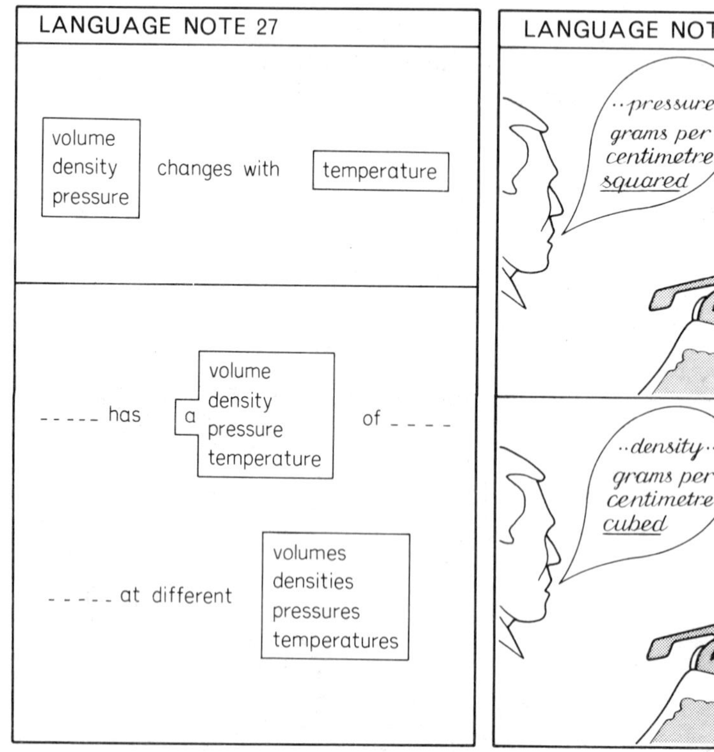

LANGUAGE NOTE 27
volume
density
pressure
changes with
temperature
----- has
a
volume
density
pressure
temperature
of ----
----- at different
volumes
densities
pressures
temperatures

LANGUAGE NOTE 28
..pressure..
grams per
centimetre
squared
BUT
g/cm²
..density..
grams per
centimetre
cubed
BUT
g/cm³

(a) density
(a) volume

a graph
a scale
a line

steam
wax

show
burn

vertical
horizontal
pure

gram(me) (g)
millilitre (ml)

approximate(ly)
exact(ly)

per (/)
times (x)

SECTION C: MOLECULES IN MOTION

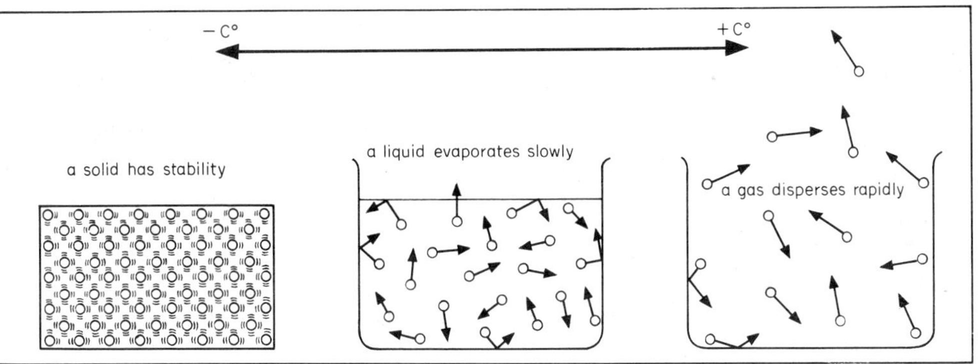

fig. (j) molecules in motion

All matter is composed of molecules. Molecules are always in motion. Temperature affects the speed of the molecules.

For example, in a solid, each molecule remains in a stable position. However, it does not remain motionless. It vibrates rapidly between the adjacent molecules, but it does not escape from them. Therefore, a solid normally has stability, *i.e.* the surface, the shape and the volume are all stable. However, under certain conditions, a solid corrodes or decomposes. For example, certain metals corrode rapidly in acids and wood decomposes slowly in water.

A liquid does not have a stable shape, but it has a definite surface. Therefore it has a definite volume. In a liquid, each molecule moves rapidly among the other molecules. It does not remain in the same position, and a molecule sometimes breaks through the surface and escapes. At normal temperatures, a liquid evaporates slowly.

In a gas, each molecule moves at high speed. It does not remain among the other molecules. In an open container, it normally disperses rapidly into the atmosphere. It has no stability. It does not have a definite shape or a stable surface. Therefore, it does not have a stable volume either.

Exercise 11 Complete this table.

	SOLIDS	LIQUIDS	GASES
SURFACE		definite but *un*stable	*in*definite and *un*stable
SHAPE			
VOLUME			

Now make six sentences from these tables.

A	solid liquid gas	has does not have	a definite shape.

In a	solid, liquid, gas,	each molecule	remains does not remain	in a stable position.

Exercise 12 Make six more sentences from these pictures and the table.

Under	certain normal	conditions,	a	solid liquid gas	corrodes decomposes disperses evaporates	slowly. rapidly.

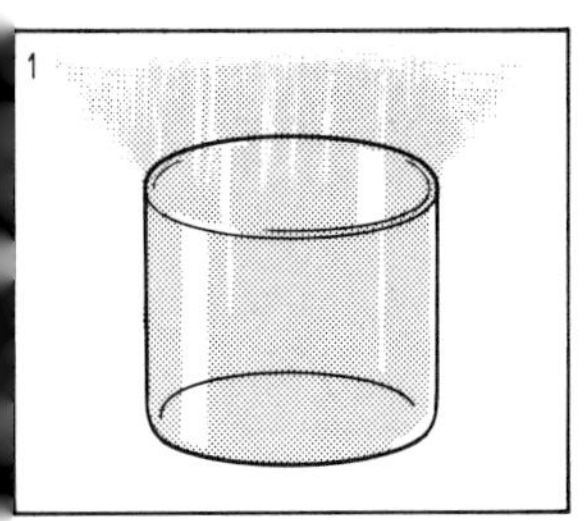

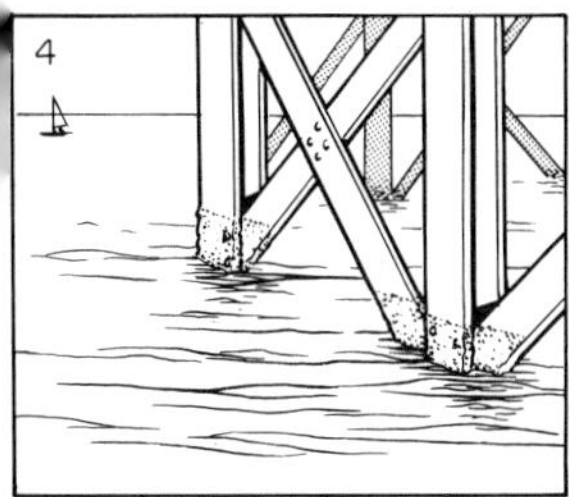

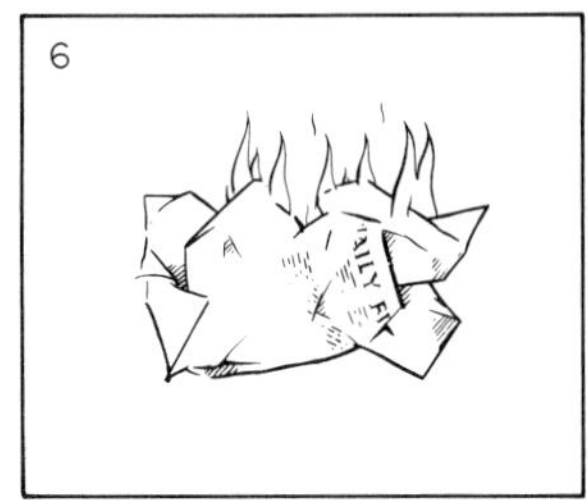

Exercise 13 Add *however* or *therefore* to each pair of sentences.

Example: Water is normally a liquid.
Above 100°C, it becomes steam.

Water is normally a liquid. *However*, above 100°C, it becomes steam.

1. In a solid, each molecule remains in a stable position.
   It does not remain motionless.
2. In a solid, each molecule remains in a stable position.
   A solid has stability.
3. A solid normally has stability.
   Under certain conditions, it corrodes or decomposes.
4. A liquid has a definite shape.
   It does not have a stable shape.
5. In a liquid, each molecule normally remains among the other molecules.
   A liquid does not disperse into the atmosphere.
6. A liquid does not disperse into the atmosphere.
   A molecule sometimes breaks through the surface and escapes.
7. A gas does not have a definite shape or a stable surface.
   It does not have a stable volume either.
8. A gas has no stability.
   A gas disperses rapidly into the atmosphere.

Exercise 14 Look at these diagrams. Then read the first two paragraphs below them. Complete the third paragraph.

hydrogen	carbon tetrachloride	benzene
H—H	Cl—C(—Cl)(—Cl)—Cl	H—C ring: six C atoms each bonded to one H, alternating single and double C–C bonds

fig. (k) molecules

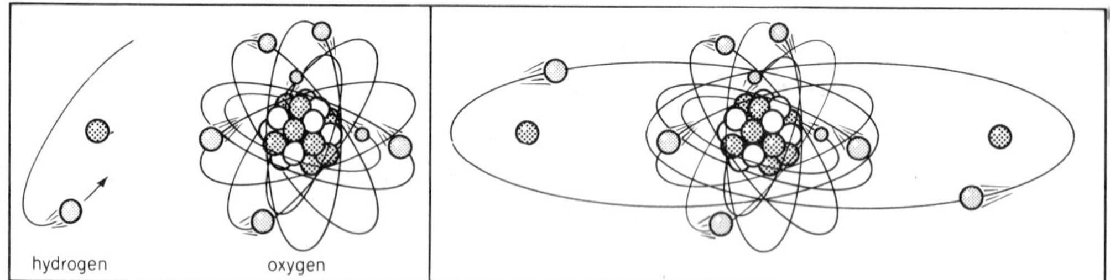

fig. (l) atoms

fig. (m) a water molecule

Element	Symbol	Compound	Formula
hydrogen	H	water	H_2O
oxygen	O	carbon tetrachloride	CCl_4
carbon	C	benzene	C_6H_6
chlorine	Cl		

fig. (n) elements and compounds

All matter is composed of molecules. All molecules are composed of *atoms*.

Some molecules have only one type of atom in them. For example, hydrogen molecules only have hydrogen atoms in them (**fig. k**). Therefore, hydrogen is an *element*. The chemical *symbol* for hydrogen is H. Other elements have different symbols. For example, the symbols for carbon and chlorine are C and Cl.

Some molecules have more than . . . . in them. For . . . . , water molecules have two . . . . atoms and one . . . . atom in them. . . . . , water is a *compound.* The chemical *formula* . . . . is H_2O. Other . . . . have . . . . example, the . . . . for . . . . and . . . . are CCl_4 and C_6H_6.

Now answer these questions.

How many hydrogen atoms are there in a hydrogen molecule?
How many carbon atoms are there in a carbon tetrachloride molecule?
How many carbon atoms are there in a benzene molecule?
How many oxygen atoms are there in carbon monoxide molecules (CO)?
How many oxygen atoms are there in carbon dioxide molecules (CO_2)?
How many oxygen atoms are there in sulphur trioxide molecules (SO_3)?

Exercise 15 Look at **figs. (a)** to **(f)** in Section A. They contain the names of twenty different elements and compounds.
Write the names etc. in the correct place in these two tables. Write them in alphabetical order. The first one in each table has been done for you.

element	normal state	symbol
aluminium	solid	Al

compound	normal state	formula
benzene	liquid	C_6H_6

Some of these compounds have elements in them which you have not written down in the first table. They are *sulphur*, *carbon* and *nitrogen*. Add them to the first table.

Common salt is also mentioned in Exercise 4. It has the chemical name *sodium chloride*. The formula is NaCl. Add this compound to the second table. Add *sodium* to the first table.

Now complete the table below. The first one has been done for you.

prefix	number of atoms	example
mon-	1	carbon monoxide (CO)
di-		(CO_2)
tri-		(SO_3)
tetra-		(CCl_4)
penta-		(N_2O_5)
hexa-		($CoCl_6$)

LANGUAGE NOTE 29	
iron corrodes	salt corrodes iron
wood decomposes	water decomposes wood
AND	
ice melts	heat melts ice
salt dissolves	water dissolves salt

LANGUAGE NOTE 30

(a) motion
(a) stability

a position
a surface

the atmosphere

adjacent
motionless
certain
open
chemical
(un)stable
(in)definite
rapid(ly)
slow(ly)

affect
remain
vibrate
escape
break
move
disperse

however
therefore

composed of

either

through
among
into

TERMINOLOGY PHYSICS AND CHEMISTRY
a molecule an atom an element a symbol a compound a formula evaporate decompose

UNIT SIX Electricity

SECTION A: ELECTROLYSIS

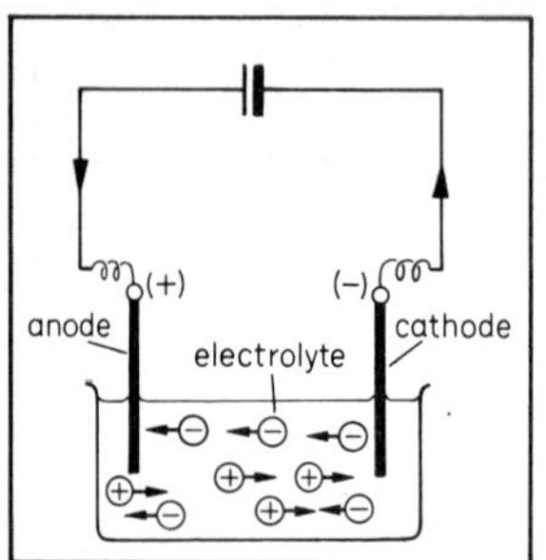

fig. (a) electrolysis

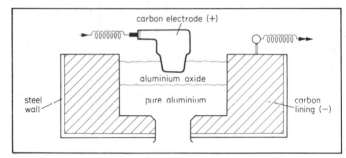

carbon electrode (+)
aluminium oxide
steel wall
pure aluminium
carbon lining (−)

fig. (b) an electric furnace

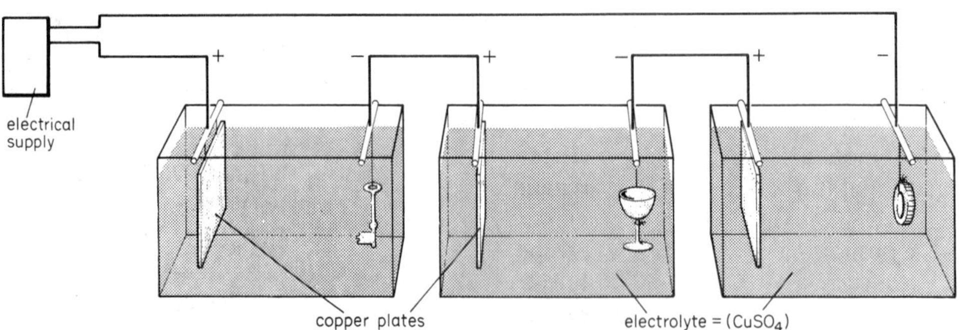

fig. (c) copper plating

The Process of Electrolysis

An *electrolytic* cell (**fig. a**) generally contains two *electrodes*. It always has a *positive* and a *negative* electrode. A positive electrode is called an *anode*, and a negative electrode is called a *cathode*. Electrolytic cells also contain an *electrolyte*. Electrolytes are usually liquids. They generally contain an acid. For example, a common electrolyte is sulphuric acid (H_2SO_4) plus water.

Electrodes are generally made of metal. They conduct *electricity* into the cell. However, carbon is sometimes used for electrodes. The electrolyte conducts an *electric current* through the cell. Under normal conditions, both sulphuric acid and water conduct electricity.

A second cell provides the current. This current travels through the anode into the electrolyte. It produces *ions* in the electrolyte. For example, both water and sulphuric acid produce positive hydrogen ions. Electrodes attract ions. A negative electrode attracts positive ions, and a positive electrode attracts negative ions. This process is called *electrolysis.*

Uses of Electrolysis

Electrolysis is used for purifying certain metals (**fig. b**).

The electric furnace contains liquid aluminium oxide. The anode is made of carbon. The furnace wall has a carbon lining. This lining is the cathode. The electric current produces negative oxygen ions in the aluminium oxide. The carbon anode attracts these ions. The pure aluminium falls to the bottom of the furnace.

Electrolysis is also used for plating metal objects (**fig. c**).

These cells contain copper salts. The anodes are copper plates. The cathodes are objects for plating. The salts provide positive copper ions. The cathodes attract these ions. The anodes dissolve slowly into the electrolyte. They provide more copper ions for plating the cathodes.

Exercise 1 Use the words *in italics in the text* to label the diagram and to complete the paragraph.

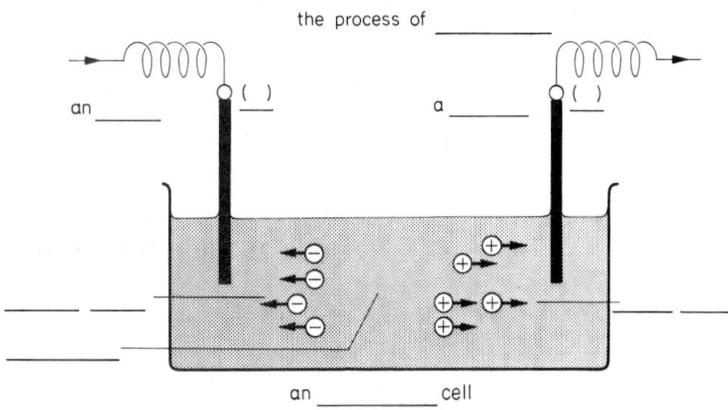

This figure shows the process of . . . . The . . . . cell contains two . . . . and a liquid . . . . The electrode on the left is the . . . . (+). The electrode on the right is the . . . . (−). An . . . . travels through the . . . . from the . . . . (+) to the

. . . . (—). The current produces . . . . in the . . . . The two . . . . attract these . . . . The . . . . (+) attracts the . . . . ions and the . . . . (—) attracts the . . . .

Exercise 2 Complete these sentences with the correct verb. Make sure the verb is in the correct form. Each verb completes *two* sentences.

attract	*contain*	*show*
conduct	*provide*	*travel*

1. An electrolytic cell generally . . . . two electrodes.
2. Carbon . . . . electricity.
3. Acids generally . . . . electricity.
4. An electric current . . . . through the electrolyte.
5. A second cell . . . . the current for the cell.
6. The electrodes . . . . the ions in the electrolyte.
7. Positive ions . . . . to the negative electrode.
8. Figs. (b) and (c) . . . . two uses of electrolysis.
9. Fig. (b) . . . . an electric furnace.
10. The carbon anode . . . . negative oxygen ions.
11. The cells in fig. (c) . . . . copper salts.
12. The metal anodes . . . . more copper ions for plating the cathodes.

Exercise 3 *Acids* and *salts* are different types of compounds. They have different *compositions*.

Examples:

ACIDS	SALTS
H_2SO_4 (sulphuric acid)	$CuSO_4$ (copper sulphate)
HCl (hydrochloric acid)	$ZnSO_4$ (zinc sulphate)
HNO_3 (nitric acid)	NaCl (common salt)

Look at the information in this table.

	Composition of	
	Acids	Salts
hydrogen molecules	all acids	no salts
oxygen molecules	some acids	some salts
molecules of a metal	no acids	all salts

Now construct six sentences with this information from the table below.

An acid *or* Acids	always	contain	. . . .
	sometimes		
A salt *or* Salts	never	contains	

Now make sentences from the pictures with the following verbs. Make sure the verb is in the correct form.

dissolve

7
Cu
Zn
Fe
Sn

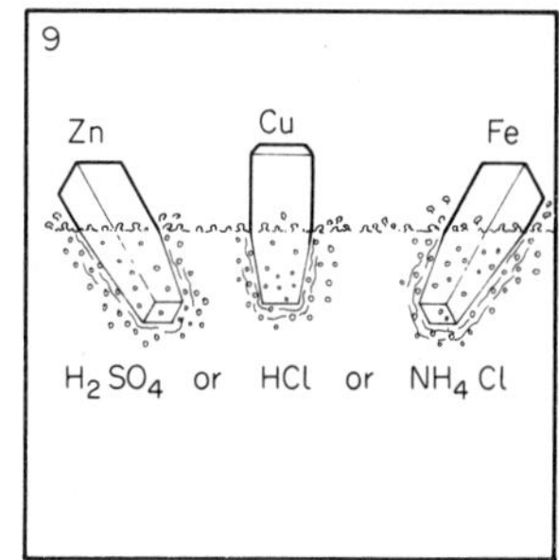

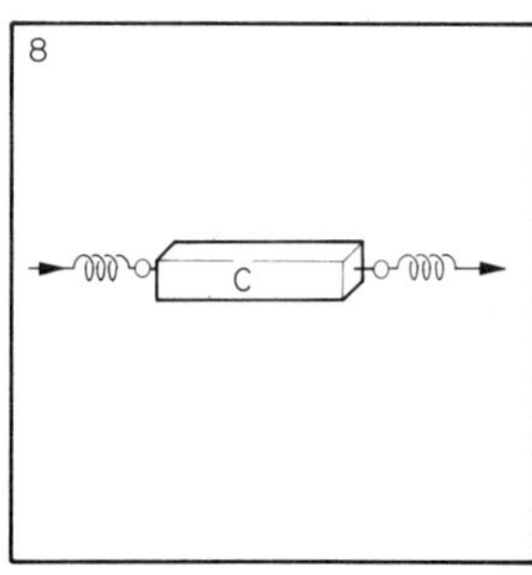

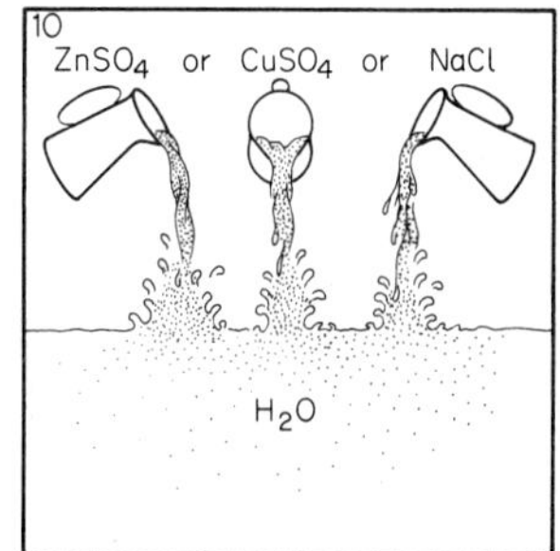

7. Metals generally . . . .
8. Carbon normally . . . .
9. Acids generally . . . .
10. Water normally . . . .

Exercise 4 Look at the picture. Complete the paragraph from the list of words.

anode	*corrode*	*plate*
attract	*however*	*produces*
called	*knives*	*sulphuric*
cathode	*lining*	*surface*
conditions	*oxide*	*therefore*

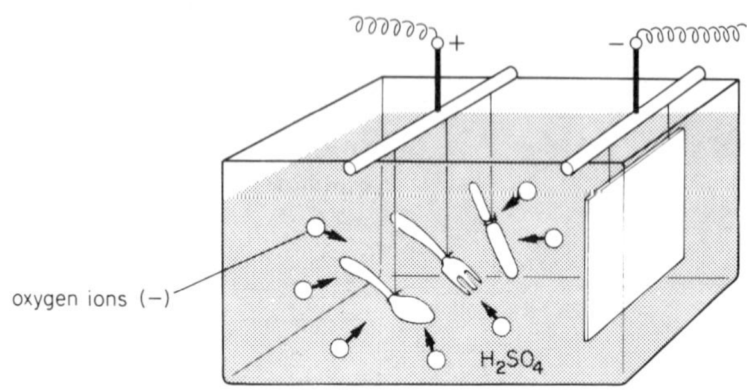

fig. (d)
the process of anodizing

Under certain . . . . , aluminium corrodes rapidly. . . . . , aluminium oxide does not normally corrode rapidly, ... electrolysis is used for plating aluminium objects with aluminium oxide. This process is . . . . *anodizing*.

In **fig. (d)**, the electrolyte contains . . . . acid. The cathode is a metal . . . . The anodes are aluminium . . . . The electric current . . . . negative oxygen ions in the electrolyte. The knives . . . . these ions. The oxygen ions and the aluminium molecules produce aluminium oxide on the . . . . of the knives. The aluminium . . . . remains on the knives and therefore they do not . . . .

Exercise 5 The *zinc* plating process and the *copper* plating process are exactly the same.

Look at the text about copper plating (p. 81) and then write eight sentences about zinc plating from the questions below. Write the sentences in the form of a paragraph.

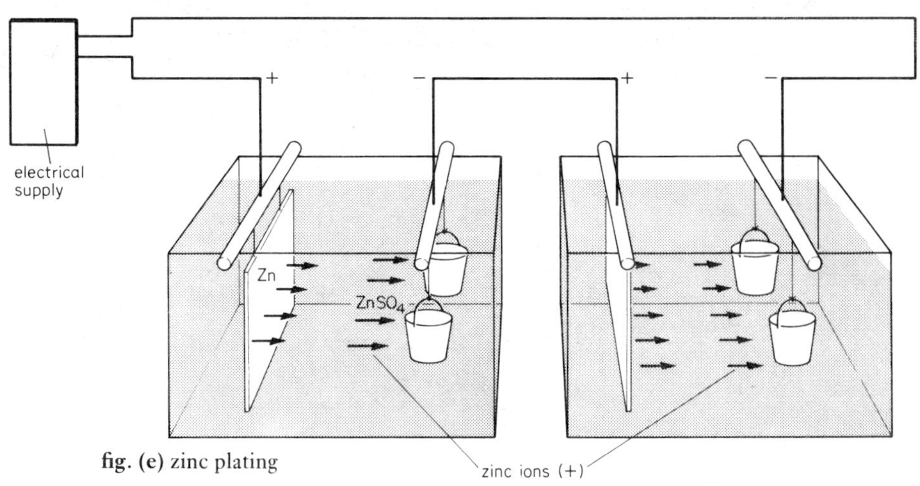

fig. (e) zinc plating

What are the electrolytic cells in fig. (e) used for?
What salt do they contain?
What are the anodes?
What are the cathodes?
What do the salts provide?
What do the cathodes attract?
What happens to the anodes?
What do the anodes provide?

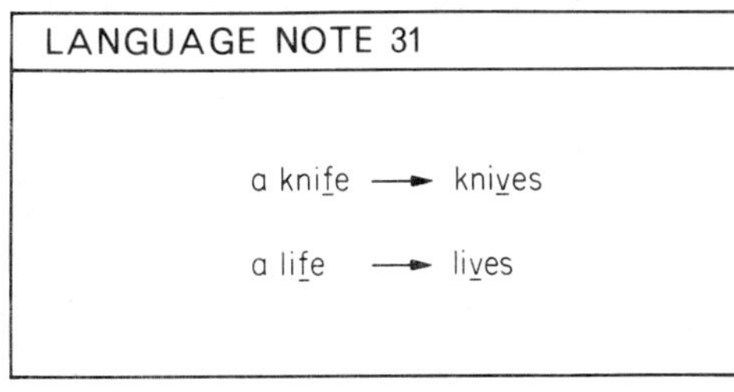

LANGUAGE NOTE 31

a knife → knives

a life → lives

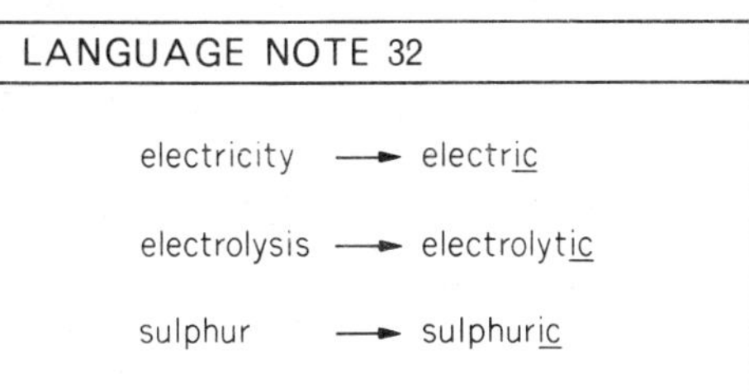

LANGUAGE NOTE 32

electricity → electric

electrolysis → electrolytic

sulphur → sulphuric

a process
a furnace
a use
a lining
a plate
an object
a salt

plating
anodizing

information
composition

contain
provide
produce
travel
purify
fall

generally

called

plus (+)

common

the bottom of . . .

ELECTRICAL TERMINOLOGY

electrolysis
an electrolytic cell
an electric current
an electrode
an anode
a cathode
an ion

positive
negative

SECTION B: BATTERIES

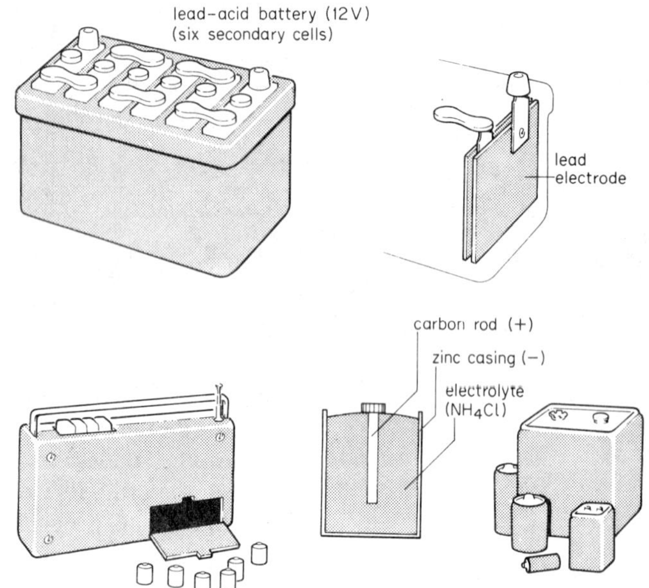

fig. (g) a lead-acid battery

fig. (f) dry cells (primary)

Electrolytic cells have several uses. Some cells are used for producing an electric current. They store an electrical *charge* in the electrolyte. *Batteries* contain this type of cell. Some batteries contain *primary* cells and others contain *secondary* cells.

Primary cells do not have a long life. It is impossible to *recharge* them. The primary cells in **fig. (f)** are called *dry cells*. These do not contain a liquid. The electrolyte is semi-solid. New cells produce approx. 1.5 V each. They generally consist of an ammonium chloride electrolyte (NH_4Cl), a carbon rod and a zinc casing. The rod is the anode and the casing is the cathode. The zinc casing corrodes with use, and the cells do not usually produce a current for more than twenty four hours.

It is possible to recharge a secondary cell many times. With careful use, secondary cells have a long life. **Fig. (g)** shows a car battery. These generally contain six lead-acid cells. These normally consist of a sulphuric acid electrolyte, lead electrodes and a plastic casing. The plastic casing does not normally corrode. Each cell produces 2V. Therefore a car battery normally produces

12 V. However, lead-acid batteries do not always consist of six cells. Some only have three. It is also possible to manufacture batteries with more than six cells for large vehicles.

Exercise 6 Are these statements true or false?

1. Batteries store an electrical charge.
2. It is possible to recharge primary cells.
3. The dry cells in fig. (f) are secondary cells.
4. In dry cells, the electrodes consist of a carbon rod and the zinc casing.
5. Dry cells contain sulphuric acid.
6. Secondary cells normally have a long life.
7. Car batteries generally contain six primary cells.
8. Lead-acid cells produce approx. 2 V each.
9. Lead-acid batteries are used in cars and other vehicles.
10. It is impossible to manufacture lead-acid batteries with more than six cells.

Exercise 7 Make sentences from this table.

A dry cell Lead-acid batteries	(does not) (do not)	. . . .	a long life. sulphuric acid. ammonium chloride. a carbon rod. lead electrodes. approx. 1.5 V. approx. 12 V. a zinc casing. a plastic casing. with use.

Exercise 8

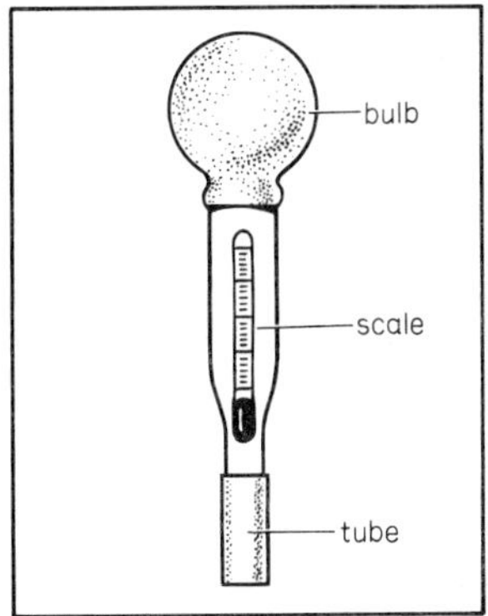

A *hydrometer* is used for checking the condition of a lead-acid battery. It measures the *specific gravity* of the electrolyte.

In a *fully charged cell*, the specific gravity of the acid is approx. 1.28.

In a *discharged cell*, the specific gravity is approx. 1.12.

Look at the pictures and make 8 instructions for checking a 12 V car battery. Use the verbs in this wordlist.

check	*press*	*release*
fill	*read*	*replace*
place		*remove*

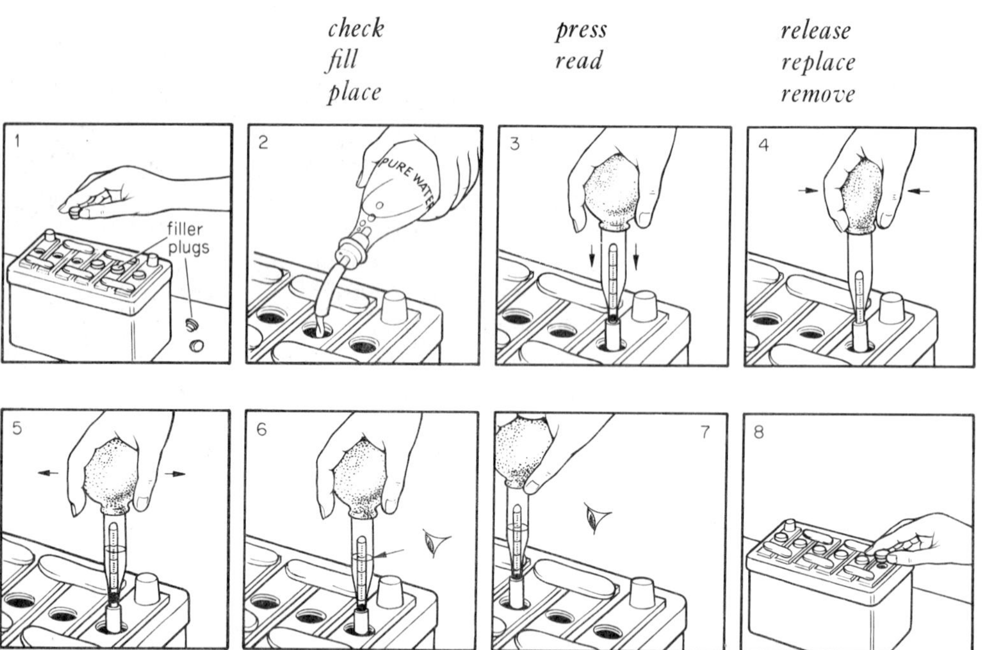

Exercise 9 With use, the composition of a lead-acid cell changes slowly. The diagram below shows the different compositions of a fully charged and discharged cell.

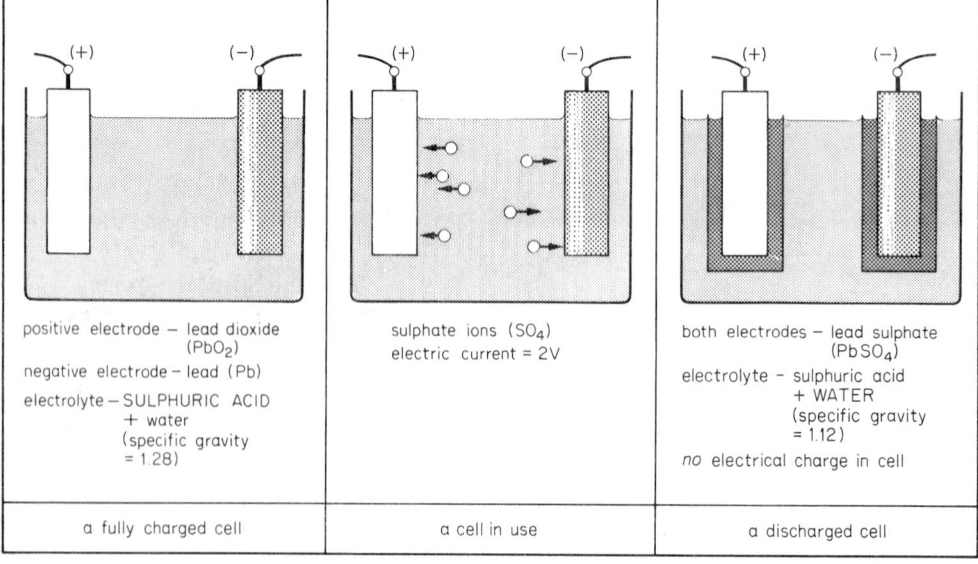

Answer the questions below and use your answers to make two paragraphs about lead-acid cells.

A fully charged cell

1. Do the electrodes consist of the same material?
2. What are the electrodes made of?
3. What does the electrolyte consist of?
4. What is the specific gravity of the cell?
5. How much current does a fully charged cell produce?

A discharged cell

1. Do the electrodes consist of the same material now?
2. What do the electrode surfaces consist of now?
3. What is the specific gravity of the electrolyte now?
4. Does the cell have an electrical charge now?

Exercise 10 Make ten sentences from this table.

It is	possible impossible	to	recharge primary cells. recharge secondary cells. make carbon electrodes. make plastic electrodes. manufacture batteries with more than six cells. store an electrical charge in a cell. store gas in an open container. use a semi-solid electrolyte in a cell. use diesel fuel in a petrol engine.

LANGUAGE NOTE 33

LANGUAGE NOTE 34

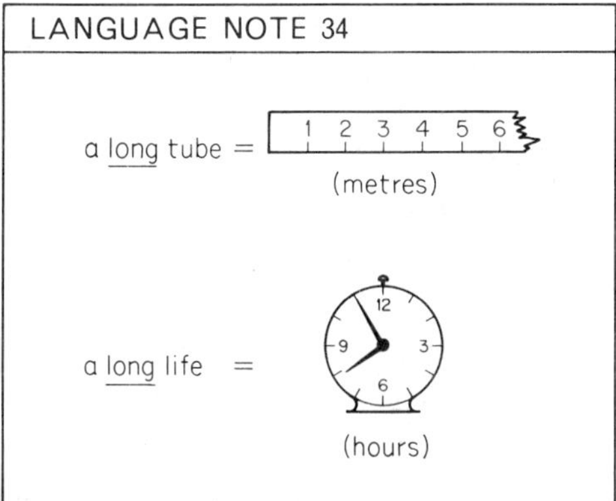

a life
a rod
an hour
a hydrometer
a bulb
a tube

store
consist (of)
manufacture
check
fill
release

several
(im)possible
dry
semi-solid
new
careful
large
discharged
full(y)

now

specific gravity

ELECTRICAL TERMINOLOGY	
a battery a charge a dry cell a volt (V) charge recharge discharge	primary secondary

SECTION C: ELECTRICAL DEVICES

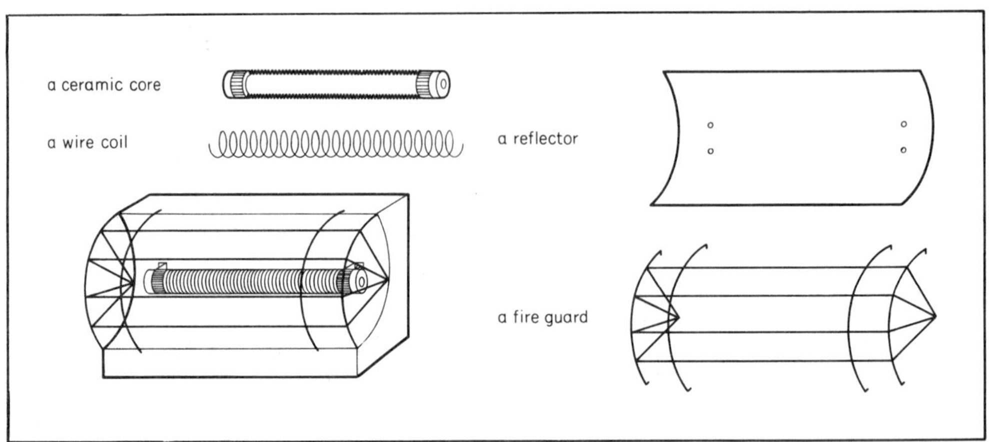

fig. (h) an electric fire

Electric fires are simple devices. Generally, they consist of elements and a reflector. The reflector is a curved sheet of metal behind the elements. It reflects heat from them. The elements radiate heat. An element consists of a coil of wire around a ceramic core. There is also a guard. This is a metal frame over the elements.

Generally, the wire coils are made of nichrome. This is an alloy with nickel and chromium in it. Nichrome is used because it has a high electrical resistance. Nichrome wire resists an electric current, *i.e.* it does not conduct electricity well. The wire converts the current into heat. The wire becomes red hot because it resists the current. However, it does not melt because nichrome has a high m.p.

The reflector is polished. Polished surfaces reflect heat well. Normally, unpolished surfaces do not reflect heat well. There is a guard over the elements because an electric fire is dangerous without it.

It is sometimes expensive to use electric fires because they require a lot of electrical power. They often require more than one kilowatt (1 kW). However, they do not usually require maintenance because they are very simple.

Exercise 11 Are these statements true or false? Convert the false statements into the negative form.

1. Nichrome wire conducts electricity well.
2. An element consists of a coil of wire around a ceramic core.
3. Nichrome wire becomes hot because it resists an electric current.
4. Nichrome wire melts because it has a high m.p.
5. Polished surfaces reflect heat well.
6. Unpolished surfaces reflect heat well.
7. Electric fires require a lot of electrical power.
8. Electric fires require maintenance.

Exercise 12 Answer these questions from the text.

1. Why is the alloy called nichrome?
2. Why is nichrome wire used in electric fires?
3. Why does an element become red hot?
4. Why does the wire not melt?
5. Why is the reflector polished?
6. Why do electric fires have a guard?
7. Why are electric fires expensive to use?
8. Why do electric fires not require maintenance?

Exercise 13 Complete the sentences from this wordlist.

around	*behind*	*over*
at	*from*	*with*
between	*into*	*without*
	of	

1. An electric fire converts an electric current . . . . heat.
2. A fire guard consists of a metal frame . . . . the elements.
3. It is dangerous to use an electric fire . . . . a guard.
4. The elements consist . . . . a ceramic core and a coil of wire.
5. There is a coil of wire . . . . the ceramic core.
6. . . . . high temperatures, the wire coil becomes red.
7. Nichrome is an alloy . . . . nickel and chromium in it.
8. There is a reflector . . . . the elements.
9. It reflects heat . . . . the elements.
10. Electric fires often require . . . . one and two kilowatts.

Exercise 14 Complete the paragraph from the wordlist.

consist	*for*	*remains*
converts	*radiate*	*simple*
devices	*reflect*	*tank*
element	*require*	*travels*
evaporate	*resistance*	*well*

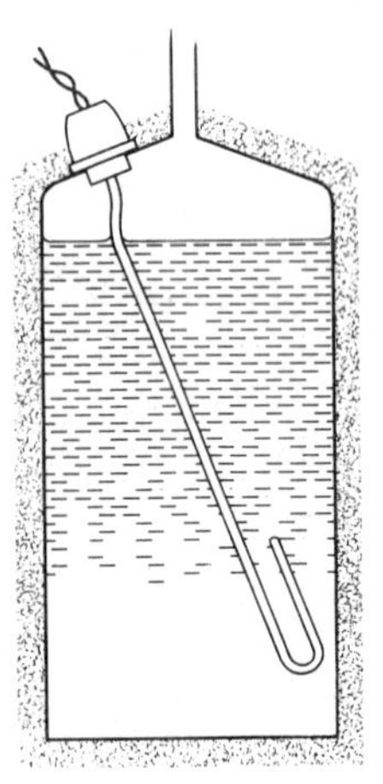

This picture shows an immersion heater. Immersion heaters are very simple . . . . They are normally used . . . . heating water. They . . . . of a long curved tube with a heating . . . . in it. The element has a high electrical . . . . It . . . . the electric current into heat. This heat . . . . through the water. Normally, an immersion heater is used in a closed . . . . Therefore, the heat . . . . in the water and the water does not . . . . These heaters do not . . . . a lot of electrical power because water stores heat . . . .

Exercise 15 Look at the picture and then answer the questions.

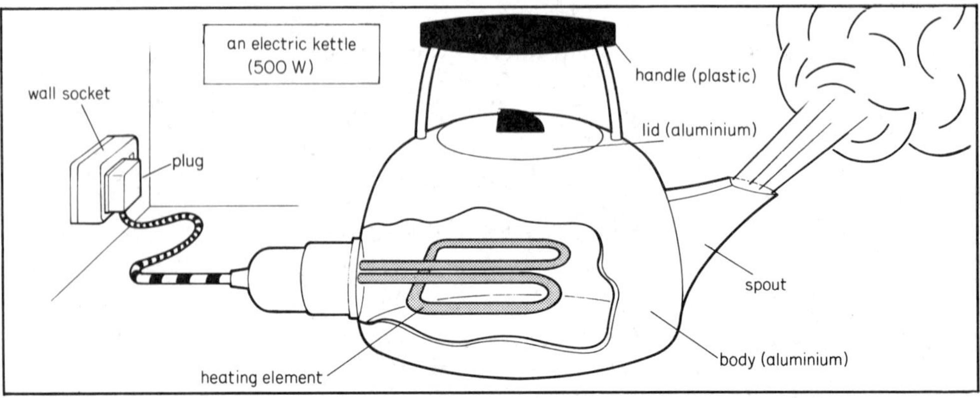

1. What is an electric kettle used for?
2. What metal is the body made of?
3. Where does the electrical power come from?
4. How much power does this kettle require?
5. Why does the element become hot?
6. Why does the kettle have a lid?
7. Why does the kettle have a spout?
8. Why does it have a plastic handle?
9. Does an electric kettle require maintenance?

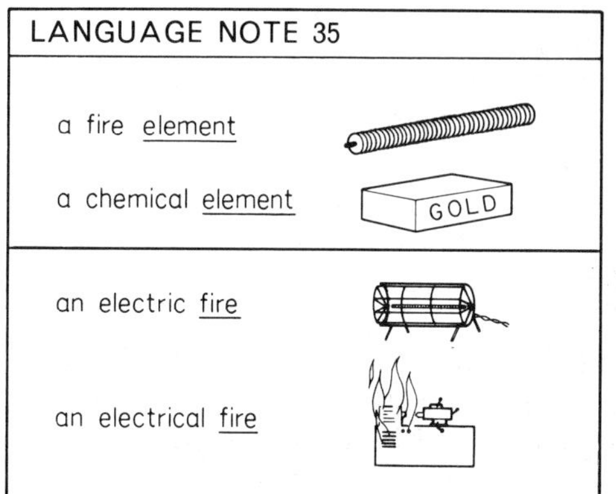

a fire
a device
a sheet
a resistance
an immersion heater
a kettle

a watt (W)
a kilowatt (kW)

wire
steel
nichrome

power
maintenance

simple
hot
(un)polished
expensive
closed

well

why . . . ?
because
behind

convert
resist
require

NAMES OF PARTS	
an element	a frame
a reflector	a socket
a coil	a plug
a core	a body
a guard	

UNIT SEVEN Properties of Engineering Materials (1)

SECTION A: FOUR PROPERTIES

It is important to know the properties of engineering materials. For example, steel is used for making girders because it is an elastic metal. Cast iron is never used for making girders because it is *brittle*. The properties of a material determine its use.

Malleability: It is easy to *roll* a *malleable* material into a new shape. A malleable material does not *fracture* easily under pressure. Gold is extremely malleable. It is possible to roll gold into very thin sheets. Copper is very malleable and so is lead. Glass is not at all malleable and nor is cast iron. It is very easy to fracture glass with a hammer. Cast iron also fractures easily.

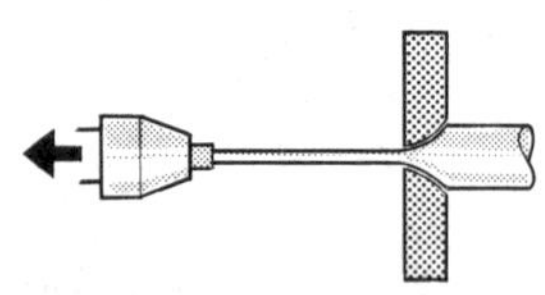

Ductility: It is easy to *draw* a *ductile* material. It does not fracture and it retains its new shape. Copper is extremely ductile. Tin is very ductile and so is aluminium. Steel is not very ductile and nor is lead. It is very difficult to draw lead into thin wire because it fractures easily.

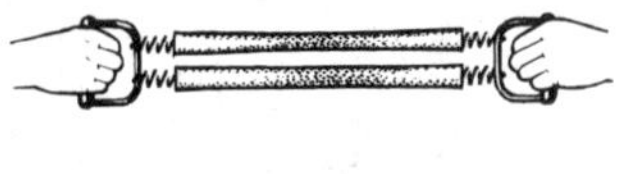

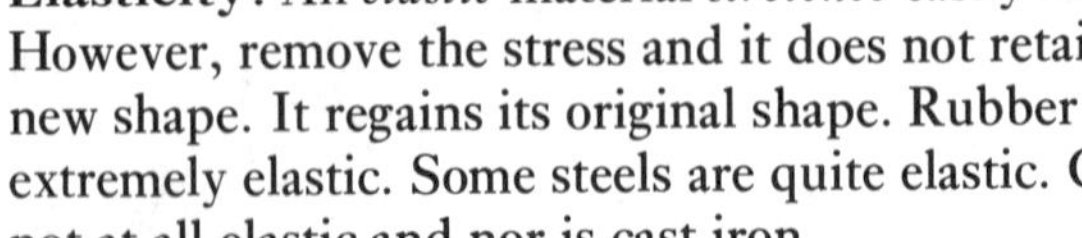

Elasticity: An *elastic* material *stretches* easily under *stress*. However, remove the stress and it does not retain its new shape. It regains its original shape. Rubber is extremely elastic. Some steels are quite elastic. Glass is not at all elastic and nor is cast iron.

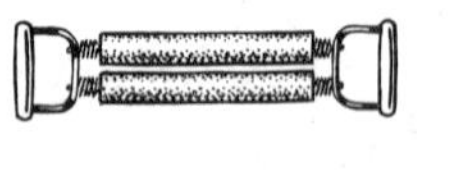

Durability: A *durable* material does not corrode easily. Under normal conditions, glass is very durable and so are plastics. Among the metals, chromium is extremely durable and so is platinum. Gold is quite durable and so is aluminium. Cast iron is not very durable and nor is steel.

Exercise 1 Complete this table from the reading text.

MATERIAL	MALLEABILITY	DUCTILITY	DURABILITY
Copper			****
aluminium	***		
gold		****	
glass		*	

extremely =	*****
very =	****
quite =	***
not very =	**
not at all =	*

Now make questions from this table and answer them.

How	malleable ductile durable	is	copper? aluminium? gold? glass?

Look at these examples.

Copper is very malleable and very durable *and* it is extremely ductile.

Aluminium is quite malleable and quite durable *and* it is very ductile.

Gold is extremely malleable and very ductile *and* it is quite durable.

Glass is not at all malleable and not at all ductile, *but* it is very durable.

Exercise 2 Now look at this table of other metals. Make sentences about these metals in the same way. Use the words *and* or *but*.

METAL	MALLEABILITY	DUCTILITY	DURABILITY
tin	****	****	****
nickel	***	****	***
cast iron	*	*	**
lead	****	**	***
steel	***	**	**
chromium	***	***	*****
zinc	****	***	****
brass	***	***	****
bronze	***	***	*****

Exercise 3 Look at the examples.

copper lead	malleable

Copper is *very* malleable *and so* is lead.

glass cast iron	ductile

Glass is *not at all* ductile *and nor* is cast iron.

Now make sentences in the same way from these words.

1. aluminium / tin — ductile
2. steel / lead — ductile
3. cast iron / glass — malleable
4. plastic / glass — durable
5. chromium / platinum — durable
6. gold / aluminium — durable
7. steel / cast iron — durable
8. bronze / brass. — malleable

Exercise 4 Look at the example. Then make sentences from the pictures in the same way.

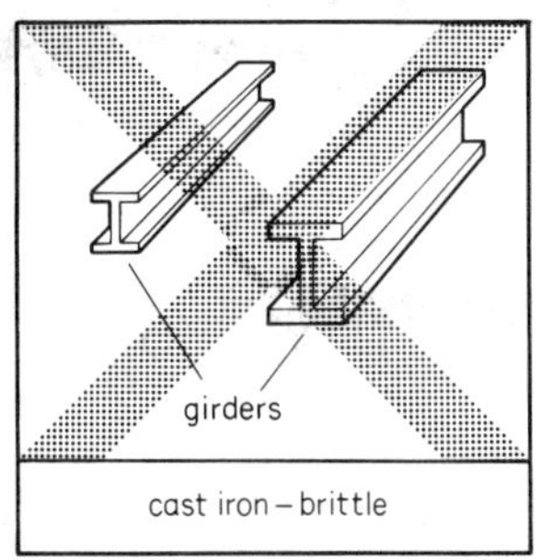

Cast iron is *not* used for making *girders* because it is *brittle*.

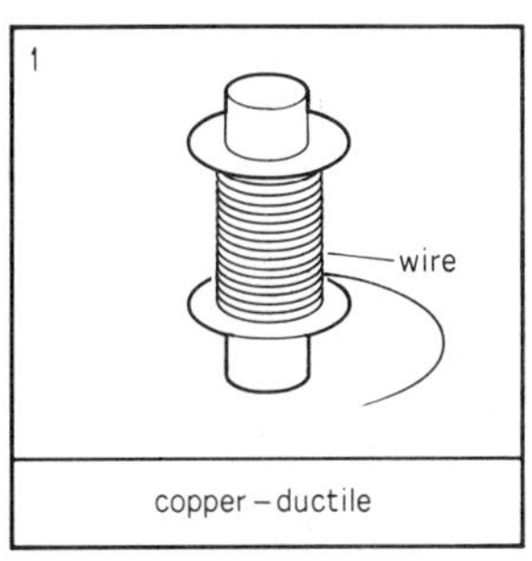

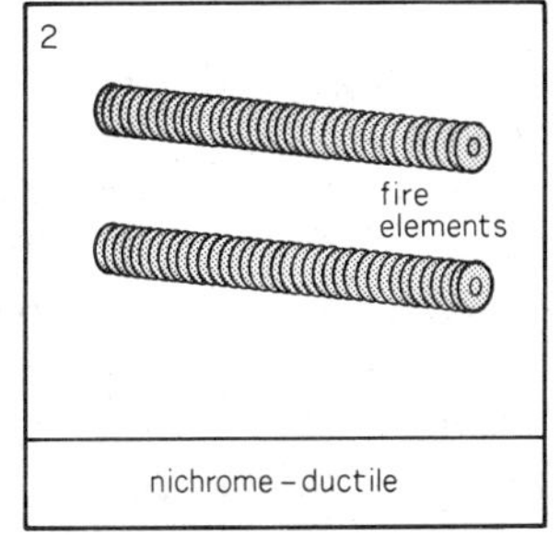

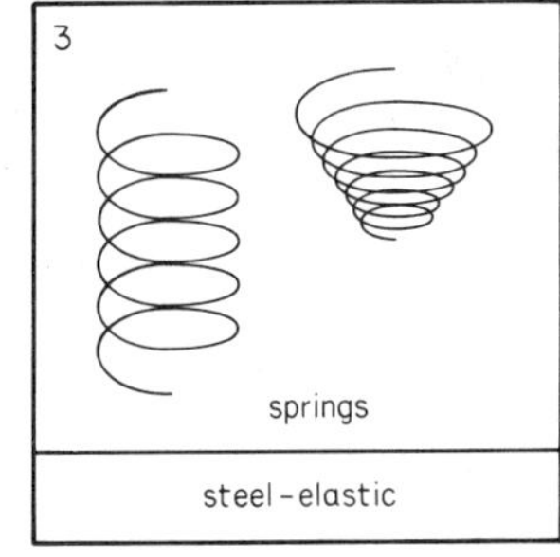

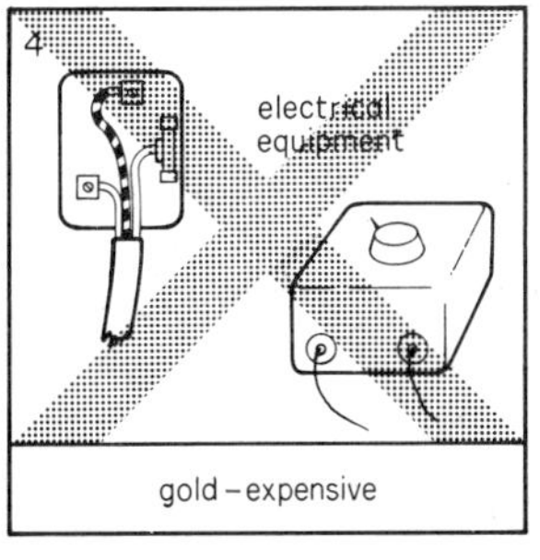

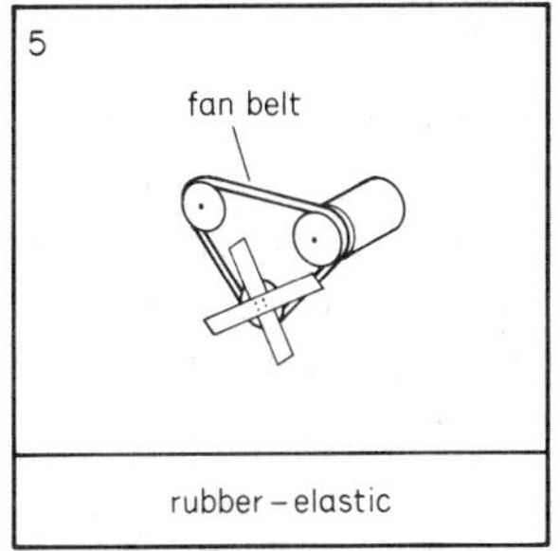

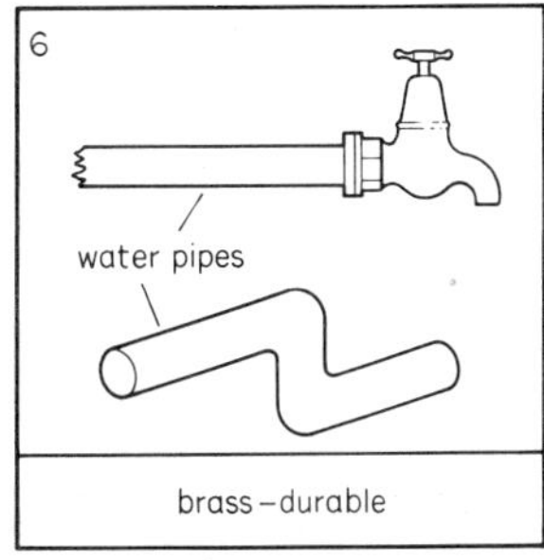

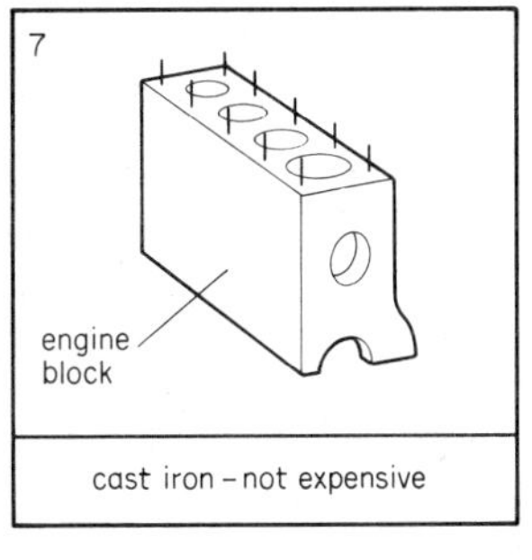

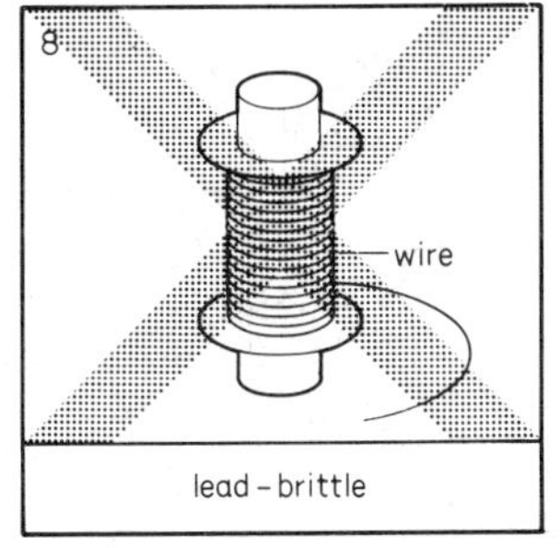

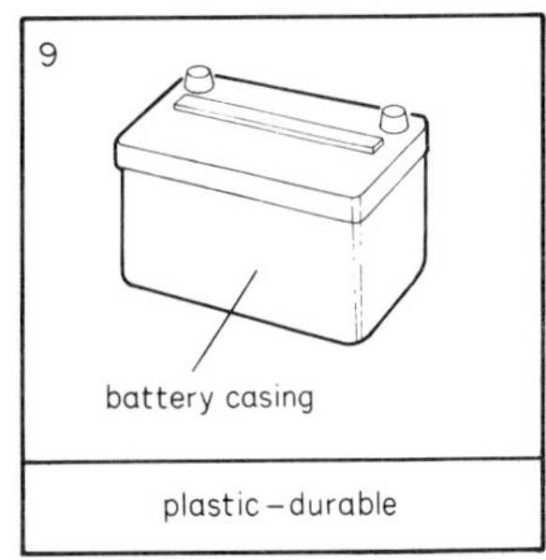

Exercise 5 Look at the pictures and complete these paragraphs from the wordlist.

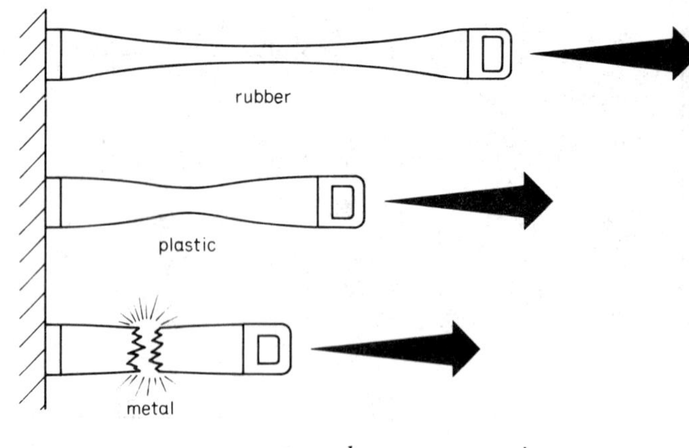

some	*extremely*	*quite*
contain	*manufacture*	*regains*
easily	*material*	*retain*
elastic	*pressure*	*stress*
elasticity	*property*	*well*

Elasticity is a . . . . of materials. An elastic . . . . stretches easily under . . . . . However, it does not . . . . its new shape. Remove the stress and it rapidly . . . . its original shape. Rubber is . . . . elastic and some plastics are also . . . . elastic.

All materials have some . . . . Among the metals, ... steels are quite elastic. These steels . . . . a little carbon. However, it is very difficult to . . . . very elastic steels because steels with a lot of carbon in them become very brittle. A brittle material fractures . . . .

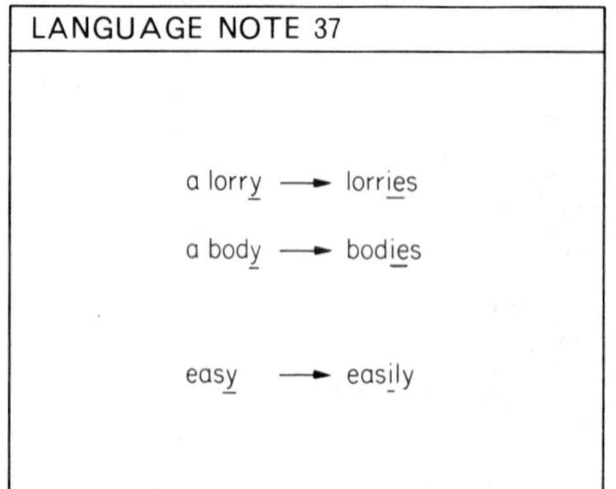

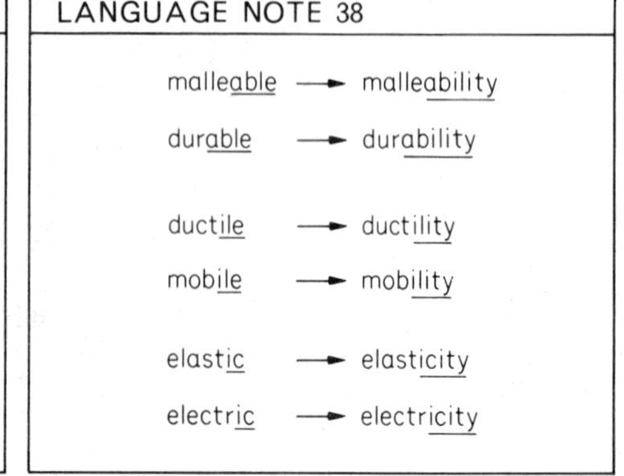

a girder
spring
a belt

engineering

know
determine
retain
regain
stretch

(and) so
(and) nor

its

important
easy (-ily)
difficult
original
thin

extremely
very
quite
not very
not at all

TERMINOLOGY (METALLURGY)	
a property	brittle
malleability (malleable)	stress
ductility (ductile)	
elasticity (elastic)	roll
durability (durable)	
	fracture
	draw

SECTION B: FERROUS METALS

	MALLEABILITY	ELASTICITY	CARBON CONTENT (approx.)
wrought iron	good	good	0.05%
low carbon steels	quite good	quite good	0.08–0.25%
medium carbon steels	not very good	not very good	0.25–0.65%
high carbon steels	poor	poor	0.65–1.50%
cast iron	very poor	very poor	2%–5%

This is a table of ferrous metals. It compares their malleability with their carbon content. It also compares their elasticity with their carbon content.

Cast iron is the least malleable of these metals. High carbon steels are not very malleable either. However, with less carbon in them, steels become more malleable. Medium carbon steels are more malleable than high carbon steels. Low carbon steels have even more malleability. Wrought iron is the most malleable of these ferrous metals.

Wrought iron is also the most elastic of these metals. Low carbon steels are less elastic because they have more carbon in them. Medium carbon steels are less elastic than low carbon steels. High carbon steels have even less elasticity because they contain even more carbon. Cast iron has the most carbon in it. Therefore, it is the least elastic.

The more carbon these metals contain, the less malleable they are. The more carbon they contain, the less elastic they are. Wrought iron contains the least carbon and it is the most elastic. Cast iron contains the most carbon and it has the least elasticity. The steels have a medium carbon content and medium elasticity.

Exercise 6 Answer these questions.

1. Which metal has the most carbon in it?
2. Which metal contains the least carbon?
3. Which metal is the least malleable?
4. Which metal has the most elasticity?
5. Which steels contain approx. 0.65–1.50% carbon?
6. Which steels are more malleable, medium or high carbon steels?
7. Which steels have less elasticity, high or low carbon steels?
8. Which properties does this table compare with carbon content?

Exercise 7 Look at the example.

wrought iron / malleability	Wrought iron contains approx. *0.05%* carbon. Therefore, it has *good* malleability.

Now make sentences from these words in the same way.

1. low carbon steels / malleability
2. cast iron / malleability
3. high carbon steels / malleability
4. medium carbon steels / malleability

Now read the next example.

wrought iron/malleable	Wrought iron has approx. *0.05%* carbon in it. Therefore, it is *very* malleable.

Make sentences from these words in the same way.

5. wrought iron/elastic
6. low carbon steels/elastic
7. medium carbon steels/elastic
8. cast iron/elastic

Exercise 8 Read the first example and then compare these ferrous metals in the same way.

wrought iron the steels	0.05% carbon 0.08–1.50% carbon
Wrought iron contains *less* carbon than the steels.	

Use the words *more* or *less*.

1. cast iron 2–5% carbon
   the steels 0.08–1.50% carbon
2. wrought iron very elastic
   cast iron not at all elastic
3. high carbon steels poor malleability
   low carbon steels quite good malleability
4. cast iron very poor elasticity
   high carbon steels poor elasticity

Now read the next example.

medium carbon steels	not very elastic
cast iron	not at all elastic
low carbon steels	quite elastic

Medium carbon steels are *more* elastic than cast iron, but low carbon steels are *even more* elastic.

Use the words *more* or *less* and *even.*

5. high carbon steels — not very malleable
   cast iron — not at all malleable
   low carbon steels — quite malleable

6. low carbon steels — 0.08–0.25% carbon
   medium carbon steels — 0.25–0.65% carbon
   wrought iron — 0.05% carbon

7. low carbon steels — quite good elasticity
   medium carbon steels — not very good elasticity
   high carbon steels — poor elasticity

Now read the next example.

the steels	malleable
wrought iron	very malleable
cast iron	not malleable

The steels are *less* malleable than wrought iron, but cast iron is the *least* malleable of these metals.

Use the words *more* and *most*, or *less* and *least*.

8. the steels — elastic
   wrought iron — very elastic
   cast iron — not elastic

9. the steels — ductile
   cast iron — not ductile
   wrought iron — very ductile

10. the steels — 0.08–1.50% carbon
    wrought iron — 0.05% carbon
    cast iron — 2–5% carbon

Exercise 9 This table shows the composition of four more steel alloys. It also compares their durability and their ductility.

	COMPOSITION		DUCTILITY	DURABILITY
Silicon steel	Silicon Carbon Iron	2% 0.1% 97.9%	not very good	quite good
Manganese steel	Manganese Carbon Iron	12% 1% 87%	good	poor
Nickel steel	Nickel Carbon Iron	3% 0.3% 96.7%	quite good	not very good
Tungsten steel	Tungsten Chromium Carbon Iron	14% 4% 0.7% 81.3%	poor	good

Examine the table carefully and then complete the following paragraphs. First, compare the carbon content of these four metals. Use *more* and *most*.

Nickel steel contains . . . . carbon than silicon steel. Tungsten steel contains even . . . . carbon. Manganese steel contains the . . . . carbon.

Next, compare their iron content in the same way.

Manganese steel . . . . tungsten steel.
Nickel steel . . . .
Silicon steel . . . .

Now compare their ductility. Use *less* and *least*.

Manganese steel has good ductility. Nickel steel is . . . . ductile than manganese steel. . . . . steel is even . . . . ductile and . . . . steel is the . . . . ductile of these alloys.

Then compare their durability in the same way.

Tungsten steel has good durability. . . . .
. . . . steel . . . .

Finally, complete this paragraph from the wordlist.

contain	*high*	*medium*
elements	*least*	*more*
ferrous	*less*	*most*
gives	*low*	*properties*

This table shows the composition of four . . . . alloys. They all . . . . iron and carbon, but they each contain one or two other . . . . . Their composition determines the . . . . of the alloys. For example, manganese steel contains . . . . carbon than the other three alloys. It is also the . . . . ductile of them. The table . . . . the iron and carbon content of each alloy. Manganese steel has a . . . . carbon content and silicon steel has a . . . . carbon content. Tungsten steel has a . . . . carbon content.

Exercise 10 Listen carefully to the teacher and write down the measurements he gives you.

Examples:

a) one point two three four millimetres (*1.234 mm*)

b) twelve point three four cubic centimetres (*12.34 cm*^{3})

c) a hundred and twenty-three point four square kilometres (*123.4 km*^{2})

d) one thousand and two hundred and thirty-four degrees Celsius (*1,234 °C*)

LANGUAGE NOTE 39

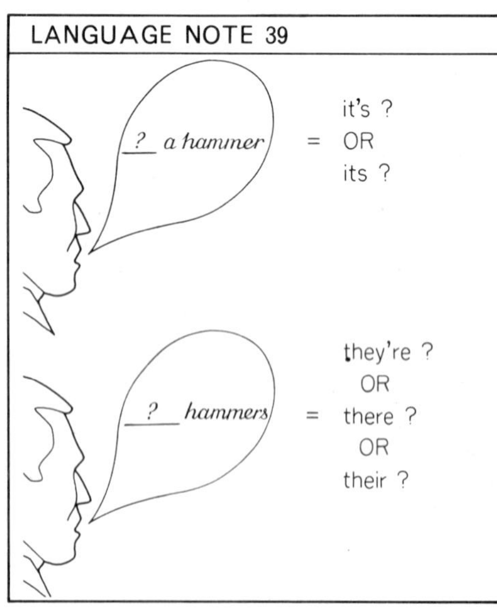

LANGUAGE NOTE 40

a high pylon (metres)
a high melting point (temperature)
a high carbon content (%)

a low bridge (metres)
a low melting point (temperature)
a low carbon content (%)

a table	compare	their
a content		
a measurement	least	ferrous
		medium
Which . . . ?	even	good
		poor

SECTION C: TENSILE STRENGTH AND HARDNESS

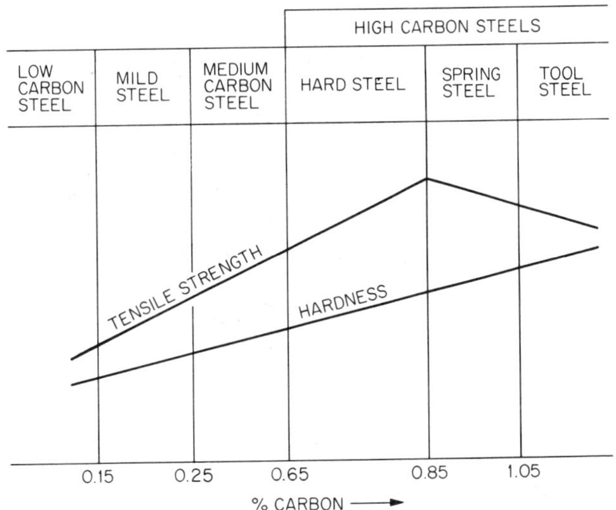

In engineering, it is important to know the tensile strength and the hardness of different alloys. The table above compares the tensile strength, the hardness and the carbon content of some common steels.

Hardness: This is the ability to withstand *abrasion.*

Low carbon steel is not very hard. It is the softest of the steels. Mild steel is harder than low carbon steel. Medium carbon steel is even harder. The high carbon steels are the hardest. Hard steel is not as hard as spring steel. Tool steel is the hardest. Among these common steels, hardness is in proportion to their carbon content. The greater their carbon content, the greater their hardness.

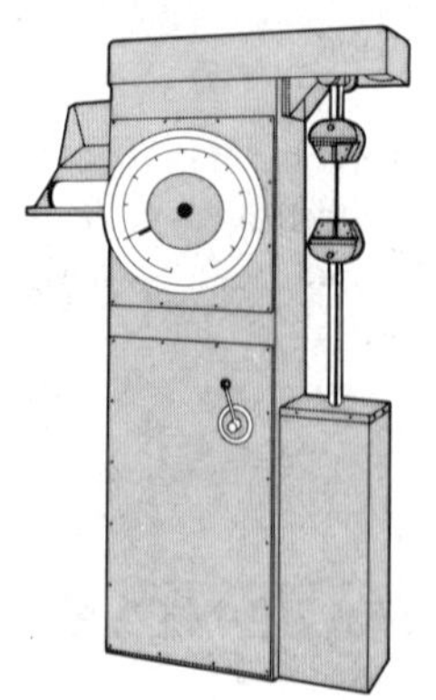
Tensile Testing Machine

Tensile strength: This is the ability to withstand *tension*.

Low carbon steel is not very strong. It is the weakest of the steels. Mild steel is stronger than low carbon steel. Medium carbon steel is even stronger. The high carbon steels are the strongest. However, their strength is not always in proportion to their carbon content. Some tool steels are not as strong as some hard steels. Below 0.85% carbon, the greater their carbon content, the stronger they are. Above 0.85% carbon, the greater their carbon content, the weaker they are.

Exercise 11 Answer these questions.

1. Which steel is the hardest?
2. Which steel is the softest?
3. Which is harder, mild steel or medium carbon steel?
4. Compare the hardness of hard steel, spring steel and tool steel.
5. Is their hardness always in proportion to their carbon content?
6. Which are the strongest steels?
7. Which is stronger, mild steel or spring steel?
8. Compare the tensile strength of mild steel, medium carbon steel and hard steel.
9. Is tensile strength always in proportion to carbon content?
10. Which steel has the greatest carbon content?

Exercise 12 Read the examples. Then make sentences from the words in the same way.

Example:

low carbon steel	soft
mild steel	?

Low carbon steel is *softer* than mild steel.
Mild steel is *harder* than low carbon steel.

1. spring steel hard
   hard steel ?
2. mild steel soft
   hard steel ?
3. hard steel strong
   mild steel ?
4. tool steel weak
   spring steel ?

Now read the next example.

Example:

<table>
<tr><td>spring steel
hard steel   hard</td></tr>
<tr><td>Spring steel is harder than hard steel.
Hard steel is not as hard as spring steel.</td></tr>
</table>

5. hard steel / mild steel — hard
6. spring steel / tool steel — strong
7. mild steel / hard steel — weak
8. low carbon steels / high carbon steels — soft

Exercise 13 Look at this example.

Example:

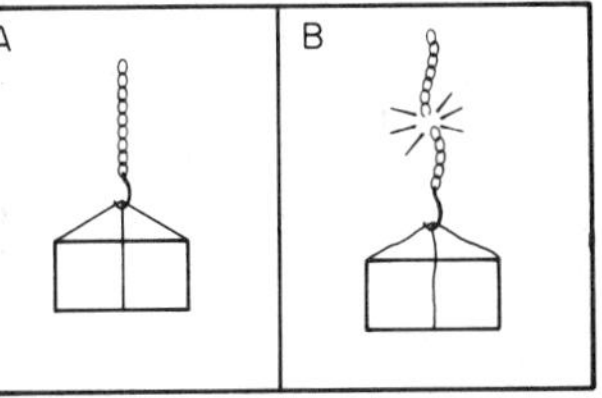

<table>
<tr><td>stronger</td><td>greater strength</td></tr>
<tr><td colspan="2">The first chain is stronger than the second chain.

The first chain has a greater strength than the second chain.</td></tr>
</table>

Make comparisons from these pictures in the same way. Choose one word or phrase from each list for each pair of pictures.

colder	*greater area*
deeper	*greater depth*
heavier	*greater height*
higher	*greater length*
hotter	*greater thickness*
larger	*greater volume*
longer	*greater weight*
thicker	*greater width*
wider	*higher temperature*
	lower temperature

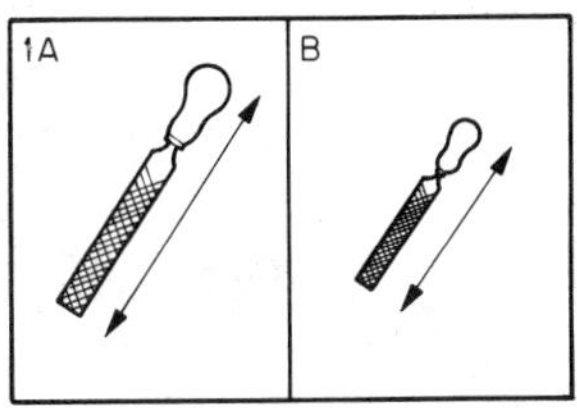

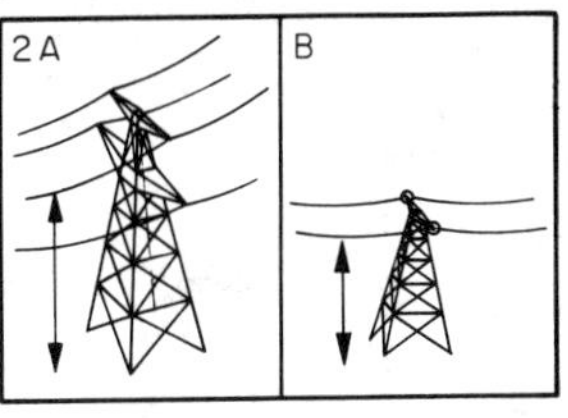

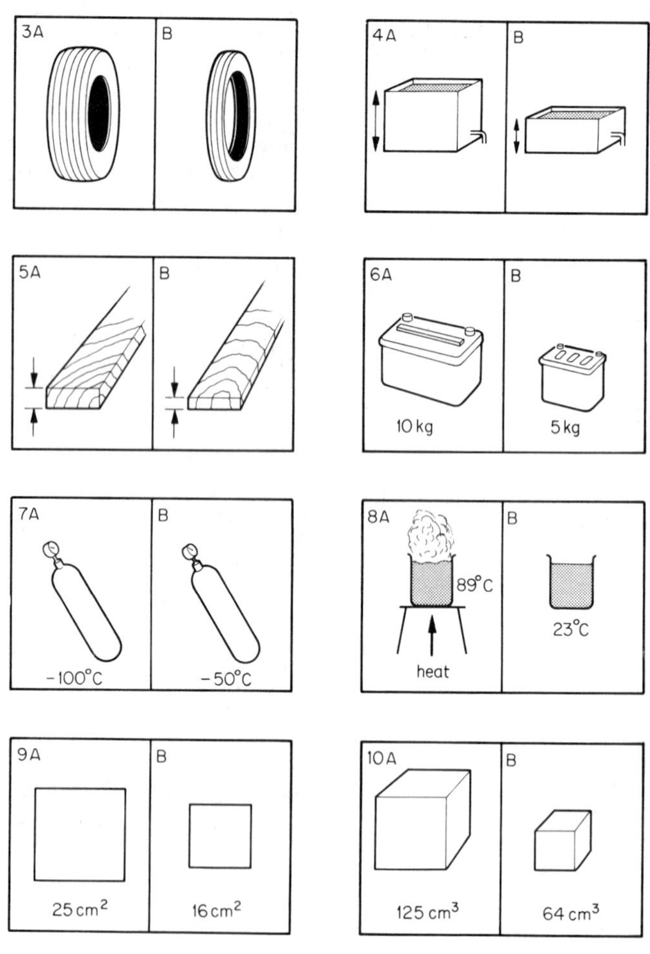

Exercise 14 Examine the diagram carefully. Then complete the paragraphs from the wordlist.

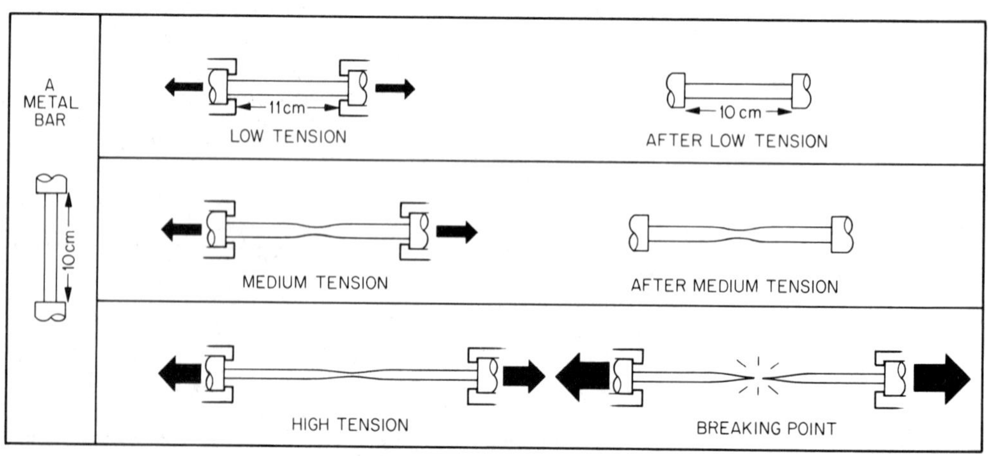

becomes	*length*	*retain*
even	*longer*	*stretches*
fractures	*original*	*tension*
impossible	*possible*	*used*
its	*regain*	*very*

A machine is . . . . for testing tensile strength. The machine slowly . . . . a metal bar under tension.

A mild steel bar is not . . . . elastic, but it stretches under low tension. Remove the . . . . and it regains its . . . . length. Increase the tension and it stretches to a greater . . . . The middle of the bar . . . . thinner. Under medium tension, the bar does not retain . . . . elasticity. Remove the tension now and it does not . . . . its original length. Under high tension, the bar becomes . . . . thinner. Finally, it . . . . in the middle. It is . . . . to measure exactly the breaking point of the bar.

Exercise 15 There are several ways to describe *properties*. First, write the adjective for each property in this table.

PROPERTY	ADJECTIVE
malleability	
ductility	
elasticity	
brittleness	

Now use the adjectives and this wordlist to complete the descriptions.

break draw roll stretch

1. It is easy to . . . . a . . . . material.
2. It is easy to . . . . a . . . . material.
3. It is easy to . . . . an . . . . material.
4. It is easy to. . . . a . . . . material.

Now look at this example.

The opposite of durable is *corrodible*.
A *corrodible* material *does not withstand* corrosion well.

Complete these descriptions from the wordlist.

break change regain withstand

5. The opposite of hard is *soft*.
   A *soft* material . . . . abrasion well.
6. The opposite of elastic is *ductile*.
   A *ductile* material . . . . its shape easily.
7. The opposite of malleable is *stiff*.
   A *stiff* material . . . . its shape easily.
8. The opposite of brittle is *tough*.
   A *tough* material . . . . easily.

Complete this table with the names of the properties.

PROPERTY	ADJECTIVE
	durable
	corrodible
	hard
	soft
	stiff
	tough

Here is another table with the names of five more properties. Write down the adjective for each property.

PROPERTY	ADJECTIVE	
tensile strength		(in tension)
compressive strength		(in compression)
refractoriness		
conductivity		
resistivity		

Finally, complete these descriptions. Look at the example first.

Tensile strength: *This is the ability to* withstand tension.

9. Compressive strength: This . . . . compression.
10. Refractoriness: This . . . . refract heat.
11. Conductivity: This . . . . conduct heat or an electric current.
12. Resistivity: . . . . an electric current.

LANGUAGE NOTE 41

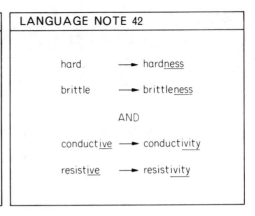

heavy → heavier

thin → thinner

hot → hotter

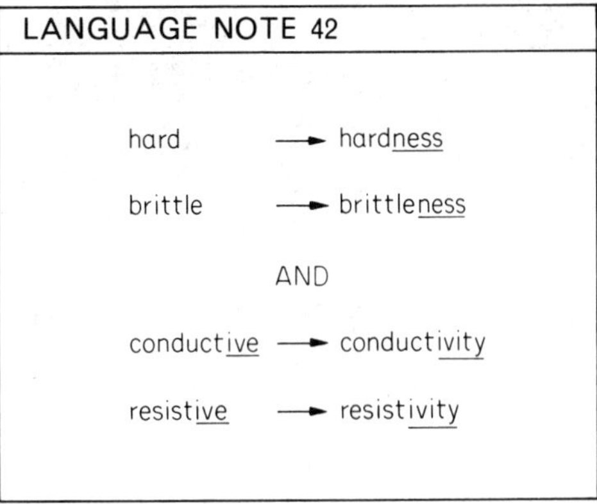

an ability	withstand	hard
a chain	test	soft
an area	increase	strong
a bar		weak
a breaking point	in proportion to . . .	great
	not as . . . as . . .	cold
	the opposite of . . .	heavy

TERMINOLOGY (METALLURGY)

tensile strength	refractoriness (refractory)
compressive strength	conductivity
hardness	resistivity
brittleness	
stiffness	tension
toughness	compression
corrodibility	abrasion

UNIT EIGHT Properties of Engineering Materials (2)

SECTION A: ALUMINIUM BRONZE

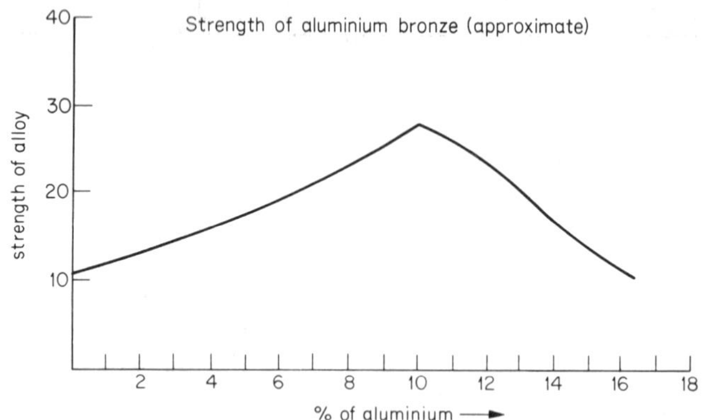

An alloy is a metallic composition with more than one element in it. For example, steel is an alloy. Common steel contains one metallic element (Fe) and one non-metallic element (C). Many other alloys only contain metallic elements. Alloys are generally stronger, more durable etc. than pure metals.

Pure aluminium is quite durable but it has poor tensile strength. Copper also has poor tensile strength. An alloy of these two metals is called aluminium bronze. Aluminium bronze has good durability and it is much stronger than either pure aluminium or pure copper.

There is always much more copper than aluminium in this alloy. Aluminium bronze with 5% aluminium (and 95% copper) is twice as strong as pure copper. With a little more aluminium in it, aluminium bronze becomes even stronger. With 10% aluminium in it, it is three times as strong as pure copper.

However, add only a little more aluminium to the alloy and it becomes much weaker. 11% aluminium bronze is a little weaker than 10% aluminium bronze. 12% aluminium bronze is not as strong as 11% aluminium bronze. 16% aluminium bronze is as weak as pure copper.

Most aluminium bronzes contain 5%–10% aluminium. A few others contain 10%–12%. Copper is much more expensive than aluminium. Therefore, the more aluminium the alloy contains, the less expensive it is.

Exercise 1 Use the phrases and words below to complete the sentences.

many	*as . . . as*
a few	*not as . . . as*
much	*twice as . . . as*
a little	*three times as . . . as*

1. Steel contains a non-metallic element, but . . . . alloys only contain metallic elements.
2. Aluminium bronze is . . . . stronger than pure copper.
3. Aluminium bronze always contains . . . . more copper than aluminium.
4. Aluminium bronze (5% Al) is . . . . strong . . . . pure copper.
5. Aluminium bronze (10% Al) is . . . . strong . . . . pure copper.
6. Aluminium bronze (11% Al) is . . . . strong . . . . 10% aluminium bronze.
7. Aluminium bronze (12% Al) is . . . . weaker than 11% aluminium bronze.
8. Aluminium bronze (16% Al) is . . . . weak . . . . pure copper.
9. Aluminium bronzes generally contain 5%–10% aluminium, but there are . . . . bronzes with 10%–12% aluminium in them.
10. Copper is . . . . more expensive than copper.

Exercise 2 Look at the examples. Then make sentences from the boxes in the same way.

Example:

copper	8.9 (g/cm^3)
aluminium	2.7 (g/cm^3)
Copper is *much heavier* than aluminium.	

Use the words *much* or *a little*.

1.

copper	$1,500 per tonne
aluminium	$450 per tonne

2.

	HARDNESS
mild steel	poor
low carbon steel	very poor

3.

	DURABILITY
tungsten steel	good
manganese steel	poor

4.

	DUCTILITY
manganese steel	good
nickel steel	quite good

Now read the next example.

Example:

	ALUMINIUM BRONZES
5–10% Al	90%
10–12% Al	10%

Most aluminium bronzes contain 5–10% aluminium. *A few* others contain 10–12%.

Use the words *most* and *a few*.

5.

	PLASTICS
heavier than H_2O	90%
lighter than H_2O	10%

6.

	PLASTICS
non-flammable	80%
flammable	20%

Now look at the last example.

Example:

BRONZES	
non-aluminium bronzes	90%
aluminium bronzes	10%

Among the bronzes, there are *many more* non-aluminium bronzes than aluminium bronzes.

Use the phrases *many more* or *a few more*.

7.

ELEMENTS	
solids	90
liquids and gases	13

8.

SOLID ELEMENTS	
metals	80
non-metals	10

9.

METALLIC ELEMENTS	
rare metals	72
common metals	9

10.

NON-METALLIC ELEMENTS	
liquids and gases	12
solids	10

Exercise 3 Look at the example. Then make eight sentences in the same way.

Example:

copper	9 g/cm^3
aluminium	3 g/cm^3

Copper is *three times as* heavy *as* aluminium.

Use these phrases.

as . . . as	*twice as . . . as*
not as . . . as	*. . . times as . . . as*
half as . . . as	

1. mercury 14 g/cm^3
   tin 7 g/cm^3
2. silver 10.5 g/cm^3
   platinum 21 g/cm^3
3. gold 19 g/cm^3
   platinum 21 g/cm^3
4. gold 19 g/cm^3
   uranium 19 g/cm^3
5. tin $9,000 per tonne
   copper $1,500 per tonne
6. copper $1,500 per tonne
   zinc $750 per tonne
7. zinc $750 per tonne
   lead $750 per tonne
8. aluminium $450 per tonne
   lead $750 per tonne

Exercise 4

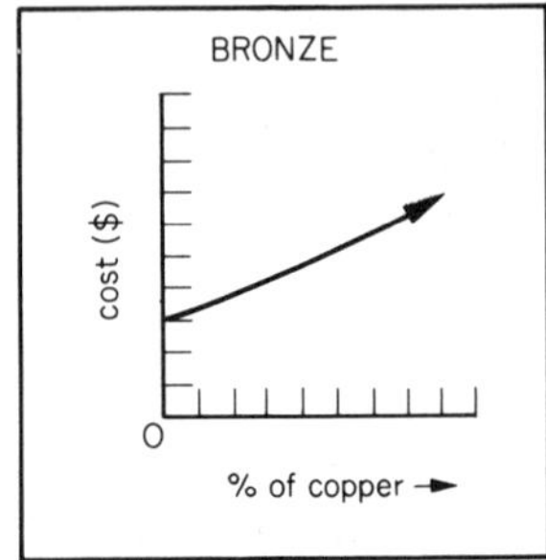

Look at the example. The *cost of aluminium* is in proportion to *the percentage of copper* in it.

The more copper the bronze contains, *the more expensive* it is.

Now describe these graphs in the same way. Use *the more* and *the less*.

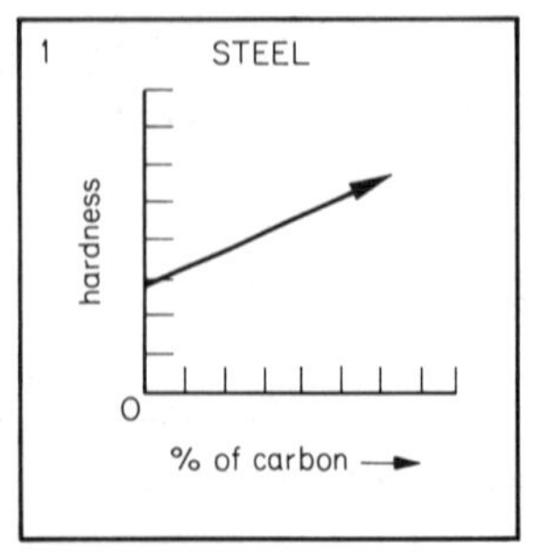

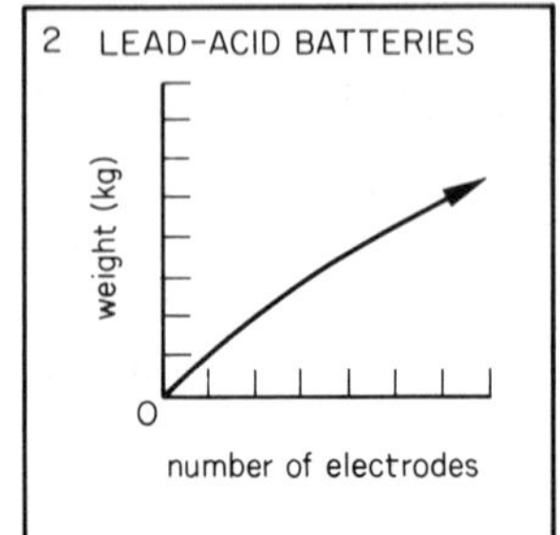

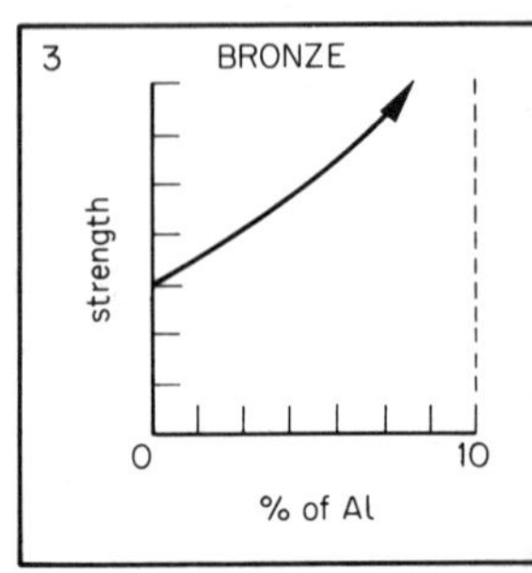

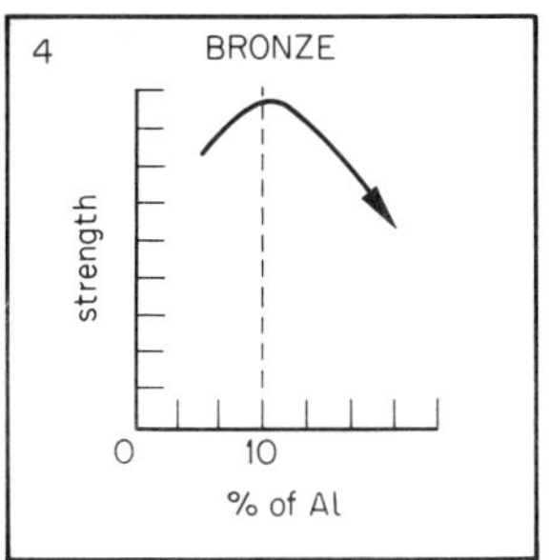

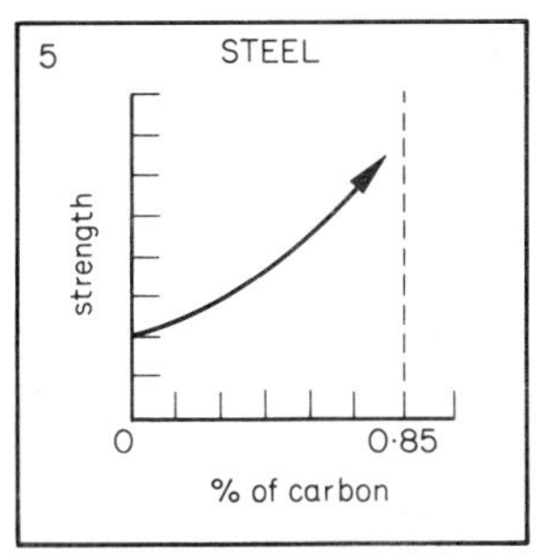

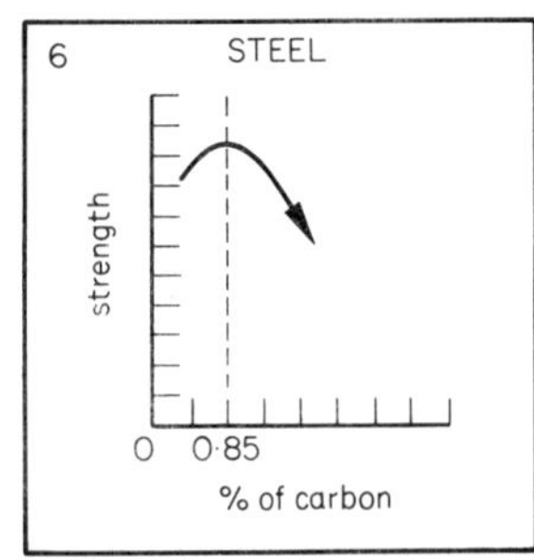

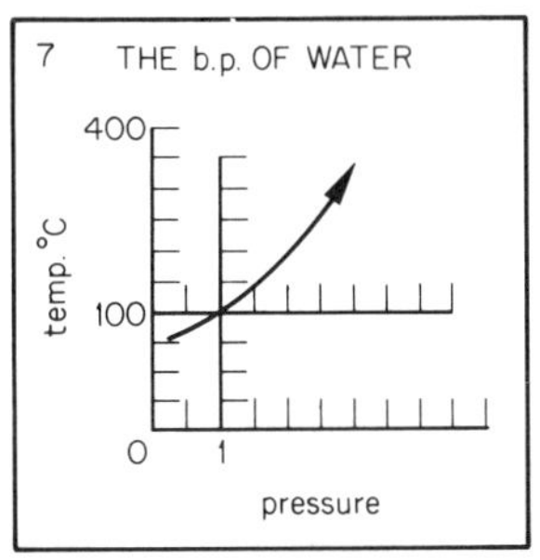

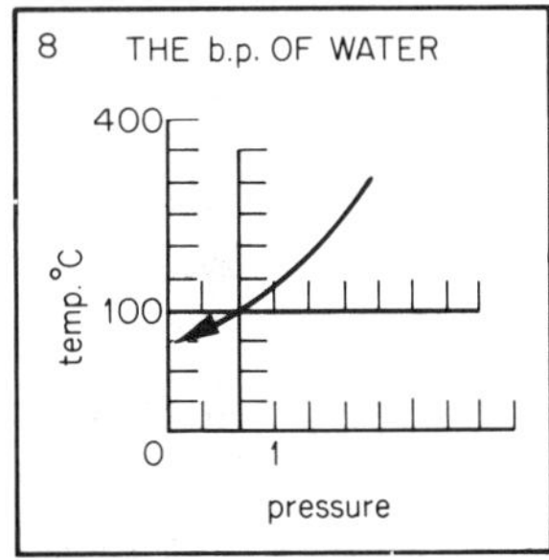

Exercise 5 Look at the pictures and then complete the paragraphs from the wordlist.

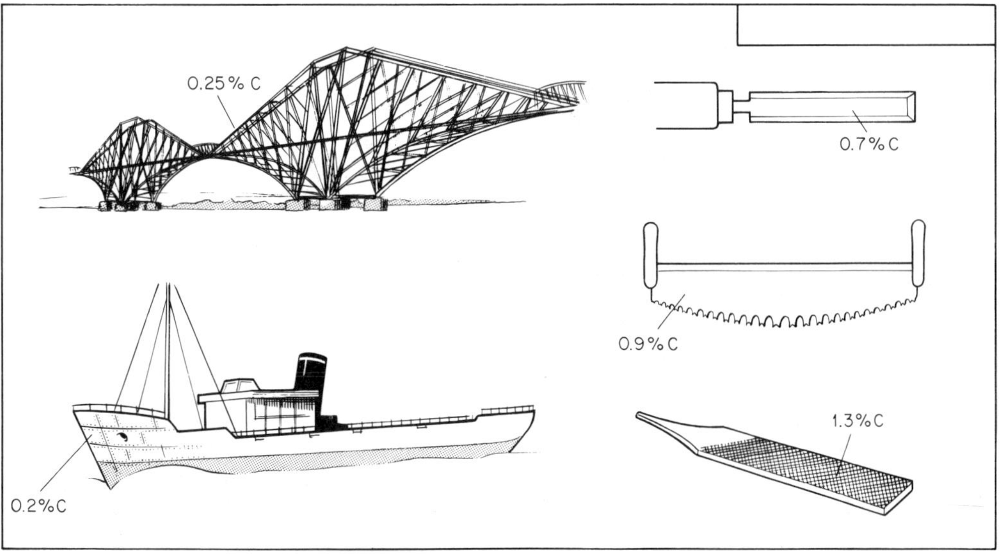

chisel	*is*	*saw*
content	*its*	*strong*
drill	*less*	*stronger*
even	*poor*	*tool*
file	*pure*	*with*
increases	*require*	*used*

Pure iron has very . . . . strength. With carbon in it, its strength . . . . rapidly. For example, 0.4% carbon steel is twice as strong as . . . . iron. However, the stronger the steel, the less ductile it . . . . 1% carbon steel is three times as . . . . as pure iron, but . . . . ductility is poor. High carbon steels are . . . . harder than steels with a medium carbon . . . . , but they have less strength.

Steels with more than 1.5% carbon are not often . . . . in engineering. For example, girders . . . . an elastic and ductile alloy. Therefore, a steel . . . . approx. 0.25% carbon in it is used. Some steel plates require even . . . . carbon in them (approx. 0.2%).
. . . . blades require a strong, hard but less elastic alloy. . . . . blades contain about 0.7% carbon. The blade of a . . . . contains about 0.9% carbon, and a . . . . blade contains more than 1% carbon.

LANGUAGE NOTE 43

Aluminium does not have good tensile strength. Copper is not very strong either.

AND

Aluminium bronze is stronger than either copper or aluminium.

a dollar ($)	add	many
a cost		(non-)metallic
twice	either . . . or . . .	light
	as . . . as . . .	rare

SECTION B: SOLDERS

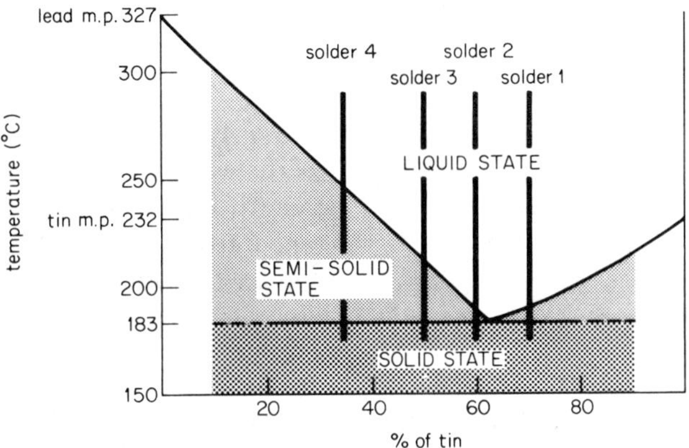

Pure metals have a definite m.p. However, many alloys do not melt at a definite temperature. They change slowly from a solid to a liquid over a range of temperatures. That is to say, they remain semi-solid between certain temperatures.

There is a range of alloys composed of lead and tin. They are called *solders*. Lead has a slightly higher m.p. (327°C) than tin (232°C). However, all solders become soft at an even lower temperature (183°C). Different solders have different uses. Their properties change with the percentage of each metal in the alloy.

In the graph, **solder 1** contains approx. 70% tin and 30% lead. Tin has a better surface than lead. That is to say, it is slightly more durable and slightly harder than lead. Therefore, this solder is the hardest and the most durable. It is also the strongest. It is frequently used for plating steel.

Solder 2 also contains considerably more tin than lead. Tin has considerably better electrical properties than lead. It also has better strength. That is why solder 2 is the best alloy for soldering electrical components. Solder 2 sets extremely quickly. That is to say, it does not remain semi-solid for a long time.

Solder 3 contains approximately as much lead as tin. It has considerably worse electrical properties than either solder 1 or solder 2. Its durability and hardness are also slightly worse. However, lead is considerably cheaper than tin. Therefore, solder 3 is used for soldering non-electrical components because it is comparatively cheap.

Solder 4 contains approx. twice as much lead as tin. It has the worst electrical properties of the four alloys. Its strength and hardness are also poor. However, it sets comparatively slowly. That is to say, it remains semi-solid over a wider range of temperatures than the other alloys. That is why it is the best alloy for soldering lead pipes.

Exercise 6 Complete the table below.

ALLOY	% OF TIN	ELECTRICAL PROPERTIES	OTHER PROPERTIES	SEMI-SOLID STATE	USE
solder 1	70%	good	hardest most durable strongest	183–190°C	plating steel
solder 2					
solder 3					
solder 4					

Look at the examples and then compare the properties of tin, lead and the solders in the same way.

Example:

solder 1 — durability
Solder 1 has the *best* durability.

Use the words *best* or *worst*.

1. solder 1 — strength
2. solder 2 — electrical properties
3. solder 4 — hardness
4. solder 4 — electrical properties

Example:

solder 2 solder 3 durability
Solder 2 has *slightly better* durability than solder 3. OR Solder 3 has *slightly worse* durability than solder 2.

Use the words *slightly* or *considerably*, *better* or *worse*.

5. lead / tin — hardness
6. lead / tin — electrical properties
7. solder 2 / solder 3 — hardness
8. solder 2 / solder 3 — electrical properties

Exercise 7 Look at the examples and then make sentences from the information in the same way.

Example:

solder 4—35% Sn 65% Pb—cheap
Solder 4 contains *considerably* more lead than tin. *That is why* it is *comparatively* cheap.

Use the words *considerably* or *slightly*, and *comparatively*.

1. solder 1 — 70% Sn 30% Pb — expensive
2. solder 3 — 48% Sn 52% Pb — soft
3. solder 2 — 60% Sn 40% Pb — strong

Example:

solder 1 — good durability — plating steel
Solder 1 has *comparatively* good durability. *That is why* it is used for plating steel.

Use the word *comparatively*.

4. solder 2 – good electrical properties — soldering electrical components.
5. solder 3 — low cost — soldering non-electrical components.
6. solder 4 — sets slowly — soldering lead pipes.

Example:

tin — better surface — more durable — lead.
Tin has a better surface. *That is to say* it is more durable than lead.

Use the phrase *that is to say.*

7. lead — 327°C — higher m.p. — tin
8. all solders — 183°C — lower m.p. — tin
9. solder 2 — semi-solid between 183°C and 188°C — sets more quickly — other solders
10. solder 4 — semi-solid between 183°C and 245°C — sets more slowly — other solders

Exercise 8 Look at the examples and then make sentences from the boxes in the same way.

Example:

solder 3	lead 52% tin 48%
Solder 3 contains approx. *as much* lead *as* tin.	

Use these phrases:

as much . . . as | *twice as much . . . as*
half as much . . . as | *. . . as much . . . as*

1. solder 4 — lead 65%, tin 35%
2. the atmosphere — nitrogen 79%, oxygen 20%
3. seawater — sodium 1%, chlorine 2%
4. seawater — calcium 0.04%, potassium 0.04%
5. cement — lime 60%, silica 20%

This diagram shows the percentages of students in the different departments of a technical college.

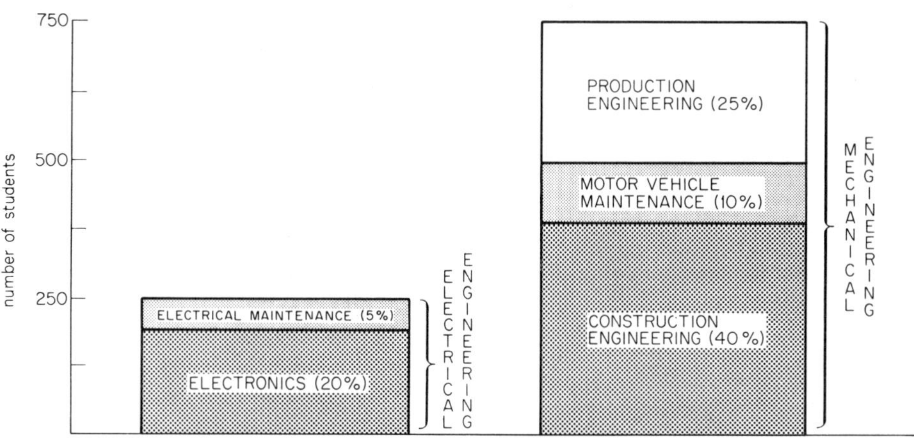

Example:

Mechanical Engineering Electrical Engineering
Mechanical Engineering has *three times as many* students *as* Electrical Engineering.

Use the phrases:

as many . . . as *twice as many . . . as*
half as many . . . as *. . . as many . . . as*

6. Electrical Maintenance
   Motor Vehicle Maintenance

7. Electrical Engineering
   Production Engineering

8. Construction Engineering
   Motor Vehicle Maintenance

9. Construction Engineering
   Electronics

10. Electronics
    Electrical Maintenance

Exercise 9 This is a table of common alloys. It gives their most important properties and their uses.

ALLOY	COMPOSITION (approx.)	IMPORTANT PROPERTIES	USES
aluminium bronze	95% Cu 5% Al	good durability golden colour	cigarette boxes etc.
duralumin	95% Al 4% Cu	light and strong	girders, tubes
nichrome	80% Ni 20% Cr	high m.p. high electrical resistance	fire elements
bronze	90% Cu 10% Sn	good durability	coins, statues
brass	67% Cu 33% Zn	good durability	ship parts
phosphor bronze	89% Cu 10% Sn (0.3% P)	hard and elastic	springs
leaded brass	66% Cu 33% Zn (1% Pb)	soft and ductile	screws, nuts and bolts
Monel metal	70% Cu 25% Ni	non-magnetic	ship parts

Now read the example. Then make paragraphs about the other alloys in the same way.

Example: Aluminium bronze *contains approx.* 19 times *as much* copper *as* aluminium. *It is used for manufacturing* cigarette boxes etc. *because it* has good durability and a golden colour.

Exercise 10 Examine the graph. Then complete the paragraph on the next page from the wordlist.

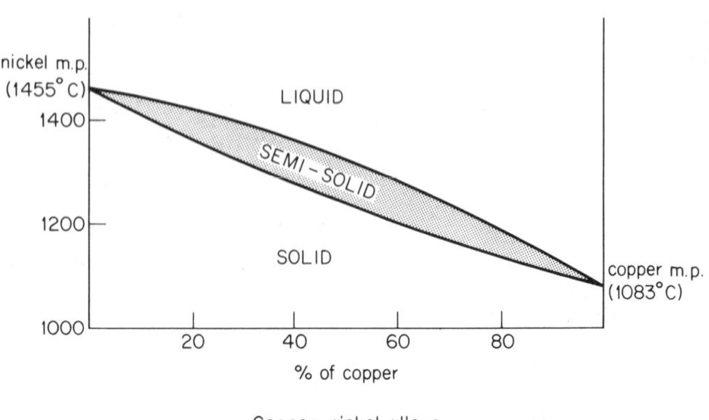

Copper-nickel alloys

becomes
between
change
definite
example
high
range
remains
slightly
slowly

Alloys do not always have a . . . . m.p. They frequently . . . . from solid to liquid over a wide . . . . of temperatures. For . . . . , the copper-nickel alloys have this property. Pure nickel has a comparatively . . . . m.p. (1,455°C) and pure copper has a . . . . lower m.p. (1,083°C). An alloy of 50% nickel and 50% copper . . . . semi-solid . . . . approx. 1,248°C and 1,312°C. That is to say, the alloy . . . . slightly soft at 1,248°C. Over the next 74°, it changes . . . . from a solid to a liquid.

LANGUAGE NOTE 44

seawater		salt water
voltmeter		gas meter
gearbox	BUT	tool box
motorcycle		motor vehicle
windscreen		wind gauge

a range
a component
a student
a colour
a cigarette
a ship
a coin
a statue

seawater
production
construction

electronics

solder
set

a long time

that is to say . . .
that is why . . .

better / best
worse / worst
cheap
mechanical
golden

frequently
slightly
considerably
comparatively
quickly

SECTION C: CONCRETE

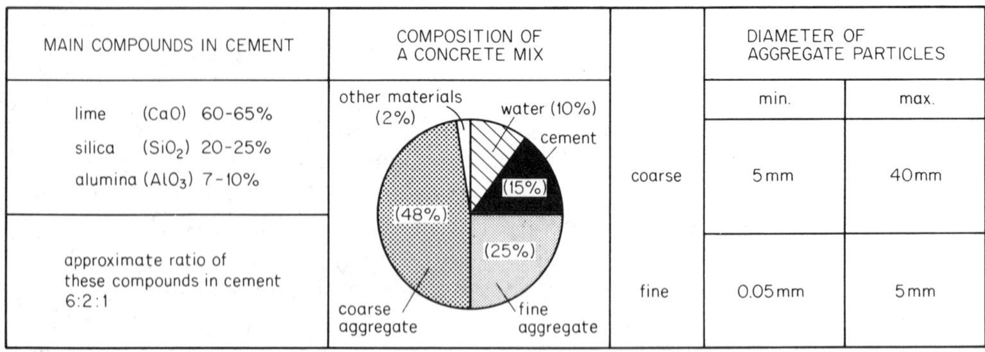

MAIN COMPOUNDS IN CEMENT	COMPOSITION OF A CONCRETE MIX		DIAMETER OF AGGREGATE PARTICLES	
			min.	max.
lime (CaO) 60-65% silica (SiO_2) 20-25% alumina (AlO_3) 7-10%	other materials (2%) water (10%) cement (15%) (48%) coarse aggregate (25%) fine aggregate	coarse	5 mm	40 mm
approximate ratio of these compounds in cement 6:2:1		fine	0.05 mm	5 mm

Concrete is a mixture of cement, aggregate and water.

Cement
This is a mixture of three main compounds. These are lime (CaO), silica (SiO_2) and alumina (AlO_3). The ratio of lime to silica is about 3:1, *i.e.* the mixture contains about three times as much lime as silica. The ratio of alumina to silica is about 1:2, that is to say, the mixture contains half as much alumina as silica.

Aggregate
Concrete contains a comparatively small amount of cement. About three quarters of a concrete mix consists of aggregate. Aggregate is composed of rock particles. *Fine* aggregate consists of sand particles with a minimum diameter of about 0.05 mm and a maximum diameter of about 5 mm. *Coarse* aggregate contains particles of more than 5 mm and less than 40 mm in diameter. Most concrete mixes contain a larger amount of coarse aggregate and a smaller amount of fine aggregate. The ratio of coarse aggregate to fine aggregate is generally about 2:1.

The Water / Cement Ratio
Their ratio affects the strength of the concrete. The normal ratio is about 2:3. Too much water decreases concrete strength considerably. A ratio of 1:1 (*i.e.* the same amount of water and cement) makes concrete too weak for most uses. It fractures too easily for construction work. A ratio of 1:3 (that is to say, too much cement) affects concrete strength in the same way.

Note: In hot conditions, concrete sometimes sets too rapidly. The mix does not retain enough water. Without enough water, concrete does not become strong enough.

Exercise 11 Are these statements true or false?

1. Lime, silica and alumina are compounds. They all contain oxygen molecules.
2. In cement, the ratio of lime to silica is about 1:3.
3. In cement, the ratio of silica to alumina is about 2:1.
4. Sand is a type of aggregate.
5. Fine aggregate consists of particles with a diameter of less than 0.05 mm and more than 5 mm.
6. Coarse aggregate consists of particles with a diameter of more than 5 mm and less than 40 mm.
7. Concrete mixes generally contain the same amount of fine and coarse aggregate.
8. Too much water in a concrete mix decreases its strength.
9. Too much cement in a concrete mix increases its strength.
10. In hot conditions, the normal water/cement ratio is sometimes not high enough.

Exercise 12 A concrete mix contains several different materials. The normal amounts for these materials are approximately:

water 10% *fine aggregate 25%*
cement 15% *coarse aggregate 48%*
other materials 2%

Look at this example of a concrete mix. Make sentences about the other mixes in the same way.

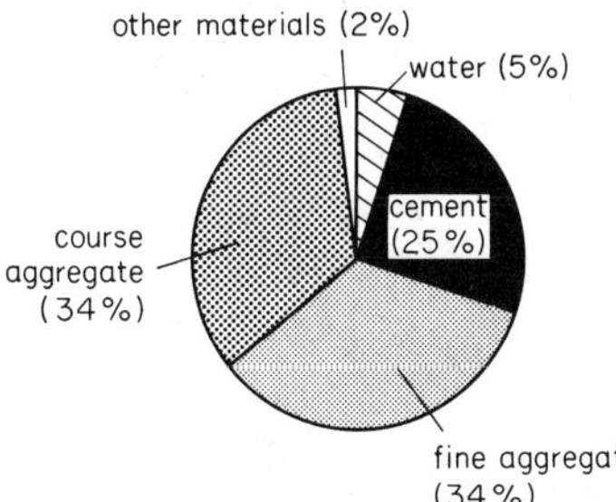

This mix contains
a small amount of water.
a large amount of cement.
a large amount of fine aggregate.
a small amount of coarse aggregate.
the same amount of coarse and fine aggregate.
the normal amount of other materials.

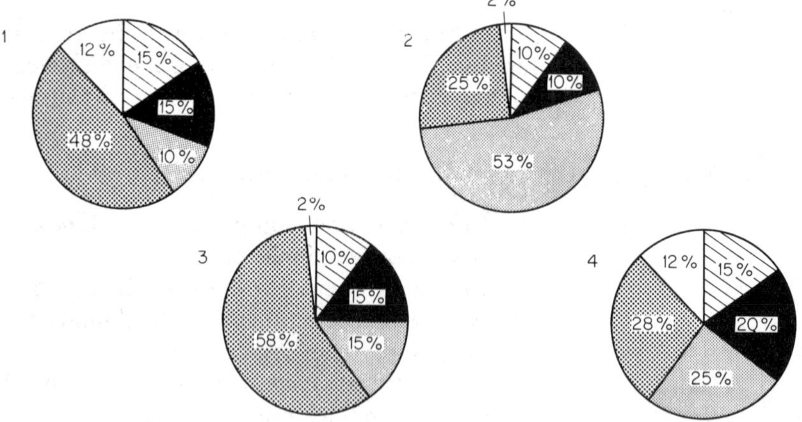

Concrete *beams* often have *rods*, *channels*, *holes* etc. in them.
Look at this example and then make sentences about the other beams in the same way.

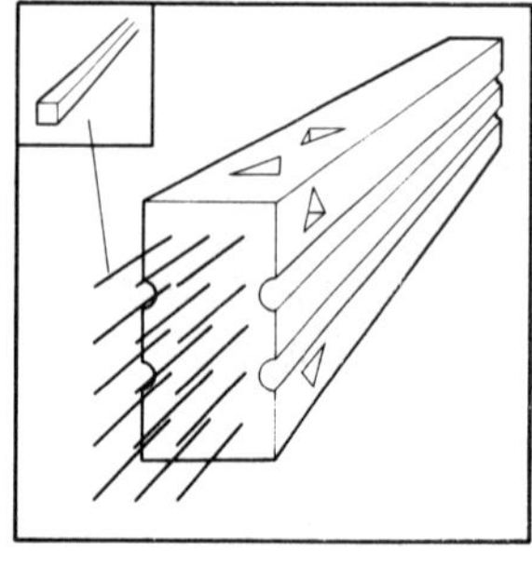

This rectangular beam has
a large number of square rods in it.
a small number of semi-circular channels in it.
a small number of triangular holes in it.

Use the words *rectangular*, *square*, *circular* etc.

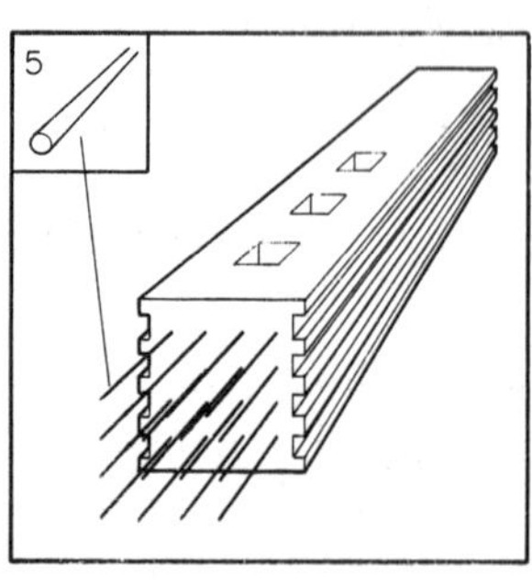

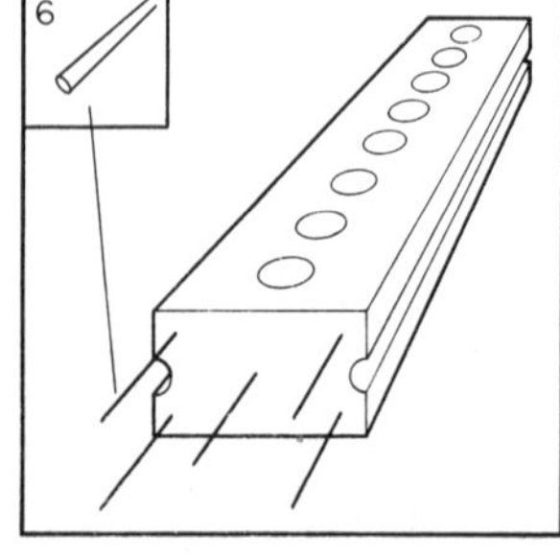

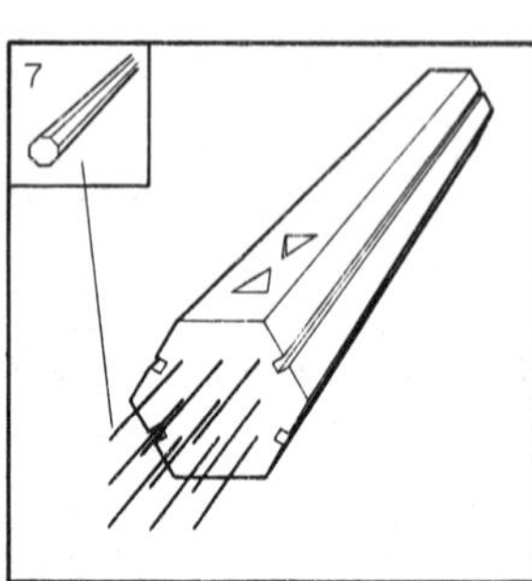

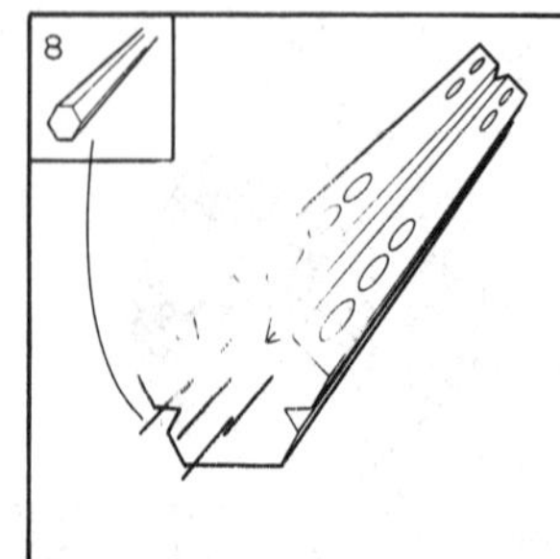

Exercise 13 Complete these sentences. Use the words in this list.

considerably	*more*	*slightly*
extremely	*less*	*too*

1. Petrol and hydrogen are . . . . flammable.
2. Sand particles with a diameter of . . . . than 0.05 mm are not used in a concrete mix because they are . . . . fine.
3. Copper has a . . . . higher m.p. (1083°C) than gold (1063°C).
4. The hotter the conditions, the . . . . rapidly water evaporates.
5. For electrical work, gold is used . . . . frequently than copper because it is very expensive.
6. Gold is . . . . heavier than aluminium.
7. High carbon steel has poor elasticity. Therefore it is . . . . brittle for construction work.
8. In liquid CO_2, the molecules move . . . . rapidly than in CO_2 gas.
9. There are . . . . fewer cars with diesel engines than with petrol engines.
10. It is impossible to use copper for making fire elements because its electrical resistance is . . . . low.

Exercise 14 Look at the example. Then make sentences from the pictures in the same way.

electrolyte
sulphuric acid/ water ratio 1:10
weak or strong?

There is *too much water* in this electrolyte.
It is *too weak*.

There is *not enough acid* in this electrolyte.
It is *not strong enough*.

Use the words *too* and *enough*.

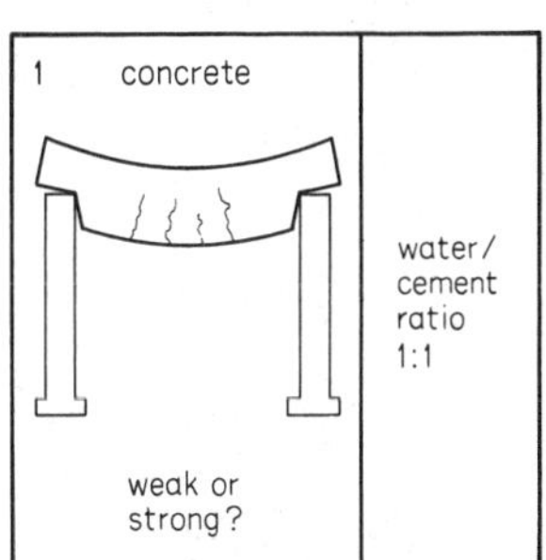

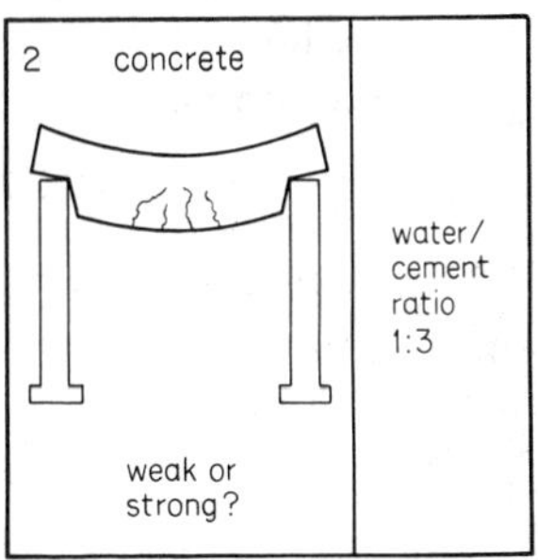

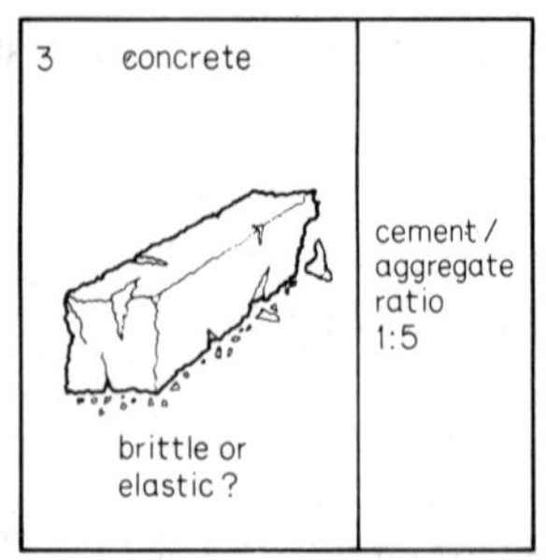

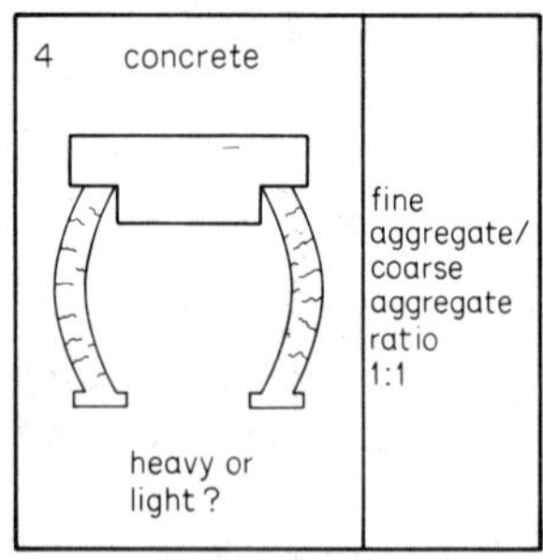

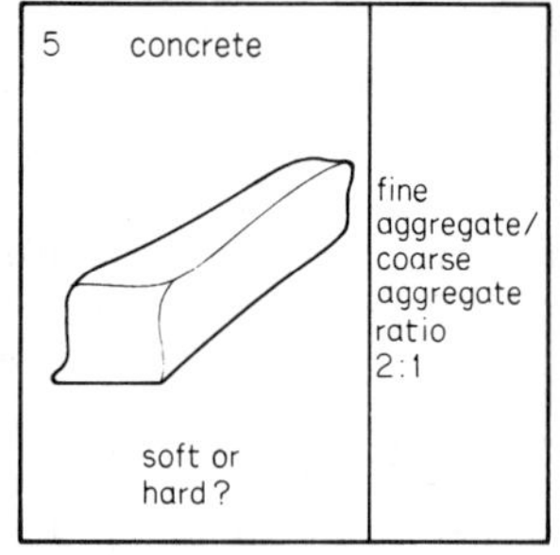

Exercise 15 Alloys, compounds etc. are composed of different amounts of pure metals, elements etc. The table on page 131 shows the different ways of describing amounts. Some of it has been done for you. Complete the table.

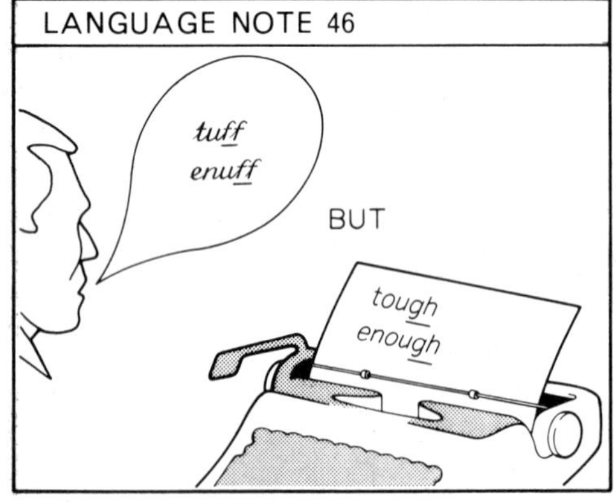

an amount
a mix
a ratio
a particle
a beam
a channel
a hole
a decimal

concrete
aggregate
rock
work

in the same way

a number of . . .

too (much)
(not) enough

small
fine
coarse
maximum (max.)
minimum (min.)

RATIOS	FRACTIONS	DECIMALS	PERCENTAGES	CIRCULAR DIAGRAMS	RECTANGULAR DIAGRAMS
1:1	*½ and ½*	*0·5 and 0·5*	*50% and 50%*	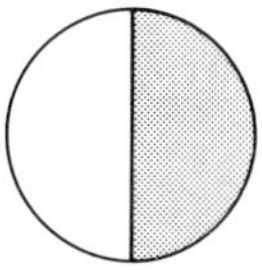	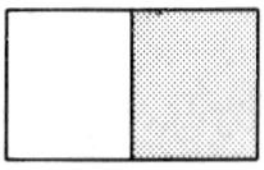
1:2	*⅓ and ⅔*			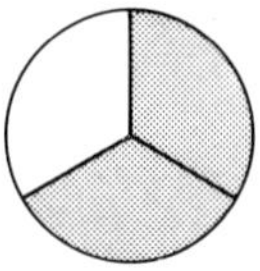	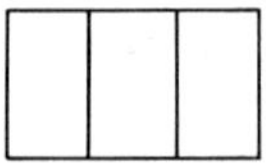
1:3	*¼ and*			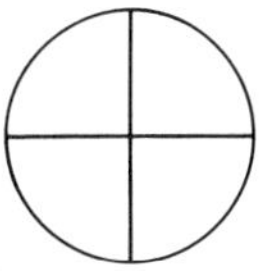	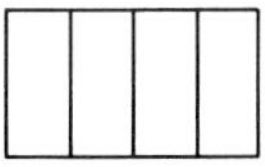
1:4			*20% and*		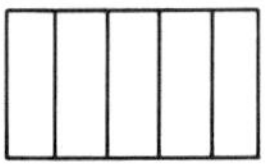
2:3			*40% and*		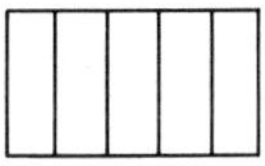
1:5		*0·167 and 0·833*		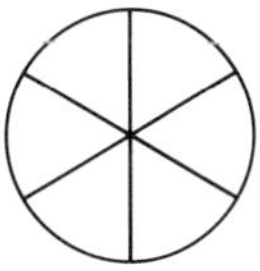	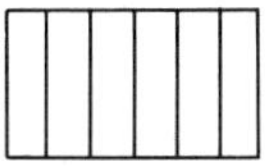
1:7		*0·125 and*		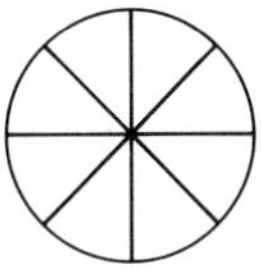	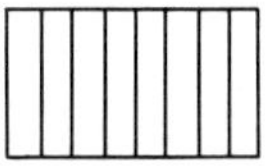
1:9				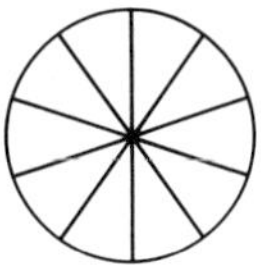	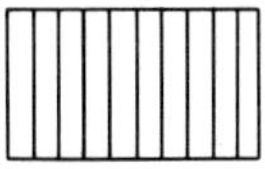

Appendix One

Abbreviations

and	&	kilometre(s)	km
approximately	approx.	kilowatt(s)	kW
boiling point	b.p.	maximum	max.
Celsius or Centigrade	°C	metre(s)	m
centimetre(s)	cm	melting point	m.p.
degree(s)	°	millilitre(s)	ml
dollar(s)	$	millimetre(s)	mm
etcetera	etc.	minimum	min.
figure	fig.	per	/
freezing point	f.p.	percent	%
gram(me)	g	temperature	temp.
id est	i.e.	volt(s)	V
kilogram(me)	kg	watt(s)	W

Appendix Two

Metals, Alloys, Non-metallic Elements and Chemical Compounds

Metals	Symbol	m.p.
aluminium	Al	660°C
chromium	Cr	1,900°C
cobalt	Co	1,490°C
copper	Cu	1,083°C
gold	Au	1,063°C
iron	Fe	1,535°C
lead	Pb	327°C
manganese	Mn	1,260°C
mercury	Hg	—39°C
nickel	Ni	1,453°C
platinum	Pt	1,769°C
silver	Ag	961°C
tin	Sn	232°C
tungsten	W	3,380°C
uranium	U	1,133°C
zinc	Zn	420°C

Alloys

brass
leaded brass

bronze
aluminium bronze
phosphor bronze

cast iron
wrought iron

duralumin
Monel metal
nichrome
solder
steel

Steels

low carbon steel
medium carbon steel
high carbon steel

mild steel
hard steel
spring steel
tool steel

manganese steel
silicon steel
nickel steel
tungsten steel

Non-metallic elements	Symbol	Normal state
bromine	Br	liquid
calcium	Ca	solid
carbon	C	solid
chlorine	Cl	gas
hydrogen	H	gas
nitrogen	N	gas
oxygen	O	gas
phosphorus	P	solid
potassium	K	solid
silicon	Si	solid
sodium	Na	solid
sulphur	S	solid

Chemical Compounds	Formula
aluminium oxide	AlO_2 or AlO_3
ammonium chloride	NH_4Cl
benzene	C_6H_6
carbon monoxide	CO
carbon dioxide	CO_2
carbon tetrachloride	CCl_4
cobalt hexachloride	$CoCl_6$
copper sulphate	$CuSO_4$
common salt (sodium chloride)	$NaCl$
hydrochloric acid	HCl
lead dioxide	PbO_2
lead sulphate	$PbSO_4$
lime	CaO
nitric acid	HNO_3

nitrogen pentoxide	N_2O_5
silica	SiO_2
sulphuric acid	H_2SO_4
sulphur dioxide	SO_2
sulphur trioxide	SO_3
zinc sulphate	$ZnSO_4$

Appendix Three

Units of Measurement in Great Britain and USA

Volume (of liquids) (GB)

2 pints = 1 quart　　1 gallon (GB) = 4.5 litres*
4 quarts = 1 gallon　　1 gallon (USA) = 3.8 litres*
1 gallon (GB) = 1.2 gallons (USA)

Length (GB & USA)

12 inches (*in*) = 1 foot (*ft*)　　1 ft = 30.5 cm*
3 feet = 1 yard (*yd*)　　1 yd = 91.5 cm*
1760 yards = 1 mile　　1 mile = 1.6 km*

Area (GB & USA)

144 square inches = 1 square foot　　1 sq. ft = $0.1\ m^2$*
9 square feet = 1 square yard　　1 sq. yd = $0.8\ m^2$*

Mass (GB)

16 ounces (*oz*) = 1 pound (*lb*)
14 pounds = 1 stone
8 stone = 1 hundredweight (*cwt*)
20 hundredweight = 1 ton
1 ton (GB) = 1.12 tons (USA)

1 oz = 28 g*
1 lb = 450 g*

1 ton (GB) = 1020 kg*
1 ton (USA) = 910 kg*

*These measurements are approximate.

Vocabulary List for Book One

GRAMMATICAL TERMINOLOGY AND ABBREVIATIONS

adjective (*adj*)
adverb (*adv*)
conjunction (*conj*)
count noun (*n.c.*)
mass noun (*n.m.*)
preposition (*prep*)
pronoun (*pron*)
verb (*v*)

The *number* and *letter* after each word refer to the *unit* and *section* in which the word first appears. Some of these words are introduced in the Teacher's Book and then appear in the wordlists at the end of each Section in the Students' Book.

ability (*n.c.*) 7C
abrasion(*n.c.* & *n.m.*) 7C
about (*prep*) 5A
about (*adv*) 3C
above (*prep*) 5A
acid (*n.c.* & *n.m.*) 1B
across (*prep*) 4B
acute (*adj*) 2B
add (*v*) 8A
adjacent (*adj*) 5C
affect (*v*) 5C
aggregate (*n.m.*) 8C
air (*n.m.*) 3C
air-cooled (*adj*) 3A
all (*adj* & *pron*) 2A
alloy (*n.c.*) 3C
also (*adv*) 1B
always (*adv*) 2B
among (*prep*) 5C
amount (*n.c.*) 8C
and (*conj*) 1A
angle (*n.c.*) 1C
anode (*n.c.*) 6A
anodizing (*n.m.*) 6A
another (*adj* & *pron*) 1B
approximate (approx.) (*adj*) 5B
approximately (*adv*) 5B
arc (*n.c.*) 4B
area (*n.c.* & *n.m.*) 7C
around (*prep*) 3A
as . . . as 7C
at (*prep*) 4A
atmosphere (*n.c.*) 5C
atom (*n.c.*) 5C

back (*n.c.*) 3A
balance (*n.c.*) 1C
bar (*n.c.*) 7C
basic (*adj*) 3C
battery (*n.c.*) 6B
beaker (*n.c.*) 1B
beam (*n.c.*) 8C
because (*conj*) 6C
become (*v*) 5A
behind (*prep*) 6C
bell (*n.c.*) 3B
below (*prep*) 5A
belt (*n.c.*) 7A
beside (*prep*) 3B
best (*adj*) 8B
better (*adj*) 8B
between (*prep*) 4A
black (*adj*) 4C
blade (*n.c.*) 1A
block (*n.c.*) 3C
body (*n.c.*) 6C
boil (*v*) 5A
boiling point (*n.c.*) 5A
bolt (*n.c.*) 1C
both (*adj* & *pron*) 3B
bottle (*n.c.*) 1B
bottom (*n.c.*) 6A
box (*n.c.*) 1B
break (*v*) 5C
breaking point (*n.c.*) 7C
brittle (*adj*) 7A
brittleness (*n.m.*) 7C
bucket (*n.c.*) 1B
bulb (*n.c.*) 6B
burn (*v*) 5B
but (*conj*) 2A

call (*v*) 4A
called 6A
can (*n.c.*) 1B
cap (*n.c.*) 1B
car (*n.c.*) 2C
careful (*adj*) 6B
case (*n.c.*) 1B
cathode (*n.c.*) 6A
cell (*n.c.*) 6A
Celsius 5A
cement (*n.m.*) 1B
Centigrade 5A
certain (*adj*) 5C
chain (*n.c.*) 7C
change (*v*) 4B
channel (*n.c.*) 8C
charge (*v*) 6B

charge (*n.c.*) 6B
cheap (*adj*) 8B
check (*v*) 6B
chemical (*adj*) 5C
chimney (*n.c.*) 2C
chisel (*n.c.*) 1A
cigarette (*n.c.*) 8B
circle (*n.c.*) 2A
circular (*adj*) 2C
close (*v*) 4B
closed (*adj*) 6C
coarse (*adj*) 8C
coil (*n.c.*) 6C
coin (*n.c.*) 8B
cold (*adj*) 7C
colour (*n.c.*) 8B
common (adj) 6A
comparatively (*adv*) 8B
compare (*v*) 7B
compass(es) (*n*) 4B
component (*n.c.*) 8B
composed (of) 5C
composition (*n.c.*) 6A
compound (*n.c.*) 5C
compression (*n.m.*) 7C
compressive (*adj*) 7C
concrete (*n.m.*) 8C
condition (*n.c.*) 5A
conduct (*v*) 5A
conductive (*adj*) 7C
conductivity (*n.m.*) 7C
considerably (*adv*) 8B
consist (of) (*v*) 6B
construction (*n.c. & n.m.*) 8B
contain (*v*) 6A
container (*n.c.*) 1B
content (*n.c.*) 7B
convert (*v*) 6C
core (*n.c.*) 6C
correct (*adj*) 4B
corrode (*v*) 5A
corrodible (*adj*) 7C
corrodibility (*n.m.*) 7C
cost (*n.c.*) 8A
cube (*n.c.*) 2C
cubic (*adj*) 2C

current (*n.c.*) 6A
curved (*adj*) 2A
curvilinear (*adj*) 2A
cylinder (*n.c.*) (i) 1B
(ii) 3A

dangerous (*adj*) 4B
decagon (*n.c.*) 2B
decimal (*n.c.*) 8C
decompose (*v*) 5C
deep (*adj*) 2C
definite (*adj*) 5C
degree (*n.c.*) 2B
density (*n.c. & n.m.*) 5B
depth (*n.c.*) 2C
determine (*v*) 7A
device (*n.c.*) 6C
diameter (*n.c.*) 4A
diagram (*n.c.*) 2B
diesel (*n.m.*) 3B
different (*adj*) 4B
difficult (*adj*) 7A
dimension (*n.c.*) 1C
discharged (*adj*) 6B
disperse (*v*) 5C
dissolve (*v*) 5A
dollar (*n.c.*) 8A
door (*n.c.*) 4A
draw (*v.*) (i) 2B
(ii) 7A
drill (*n.c.*) 1A
drum (*n.c.*) 1B
dry (*adj*) 6B
ductile (*adj*) 7A
ductility (*n.m.*) 7A
durability (*n.m.*) 7A
durable (*adj*) 7A

each (*adj & pron*) 4B
easily (*adv*) 7A
easy (*adj*) 7A
either (*adv*) 5C
either . . . or (*conj*) 8A
elastic (*adj*) 7A
elasticity (*n.m.*) 7A
electric (*adj*) 6A
electrical (*adj*) 4B

electricity (*n.m.*) 5A
electrode (*n.c.*) 6A
electrolysis (*n.m.*) 6A
electrolytic (*adj*) 6A
electronics (*n.m.*) 8C
element (*n.c.*) (i) 5C
(ii) 6C
ellipse (*n.c.*) 2A
elliptical (*adj*) 2C
engine (*n.c.*) 3A
engineering (*n.m.*) 7A
enough (*adv*) 8C
equal (*adj*) 2A
equipment (*n.m.*) 4B
escape (*v*) 5C
etcetera (etc.) 4B
evaporate (*v*) 5C
even (*adv*) 7B
exactly (*adv*) 5B
expand (*v*) 5A
expensive (*adj*) 6C
extinguish (*v*) 4A
extinguisher (*n.c.*) 4A
extremely (*adv*) 7A

face (*n.c.*) 2C
fact (*n.c.*) 5A
fall (*v*) 6A
fan (*n.c.*) 3A
feeler gauge (*n.c.*) 1C
ferrous (*adj*) 7B
few (*a* few) (*adj & pron*) 3C
figure (fig.) (*n.c.*) 5A
file (*n.c.*) 1A
file (*v*) 1C
fill (*v*) 6B
fin (*n.c.*) 3A
fine (*adj*) 8C
fire (*n.c.*) (i) 4A
(ii) 6C
(in)flammable (*adj*) 4B
flap (*n.c.*) 1B
foam (*n.m.*) 4B
formula (*n.c.*) 5C
fraction (*n.c.*) 3C
fracture (*v*) 7A
frame (*n.c.*) 6C

freeze (*v.*) 5A
freezing point (*n.c.*) 5A
frequently (*adv*) 8B
from (*prep*) 4A
front (*n.c.*) 3A
fuel (*n.c. & n.m.*) 3B
fully (*adv*) 6B
furnace (*n.c.*) 6A

gap (*n.c.*) 1C
gas (*n.c. & n.m.*) 1B
gauge (*n.c.*) 1C
gear (*n.c.*) 3C
gearbox (*n.c.*) 3C
generally (*adv*) 6A
geometrical (*adj*) 2A
girder (*n.c.*) 7A
give (*v.*) 5A
glass (*n.m.*) 1B
golden (*adj*) 8B
good (*adj*) 7B
gram(me) (g) (*n.c.*) 5B
graph (*n.c.*) 5B
great (*adj*) 7C
ground (*n.m.*) 4B
guard (*n.c.*) 4C

hacksaw (*n.c.*) 1A
hammer (*n.c.*) 1A
hammer (*v*) 1C
hand (*n.c.*) 4C
handle (*n.c.*) 1A
handlebars (*n*) 3B
handsaw (*n.c.*) 1A
hard (*adj*) 7C
hardness (*n.m.*) 7C
head (*n.c.*) 1A
heat (*n.m.*) 5A
heavy (*adj*) 7C
height (*n.c. & n.m.*) 2C
hexagon (*n.c.*) 2B
hexagonal (*adj*) 2C
high (*adj*) 2C
hoist (*n.c.*) 3B
hold (*v.*) IC
hole (*n.c.*) 8C
horizontal (*adj*) 5B
horn (*n.c.*) 4C
hose (*n.c.*) 4C
hot (*adj*) 6C
hour (*n.c.*) 6B
how. . ? (*adv*) 2C
however (*conj*) 5C
how many . . ? 2B
how much . . ? 3C
hub (*n.c.*) 4B
hydrometer (*n.c.*) 6B

ice (*n.m.*) 5A
i.e. 5A
immersion heater (*n.c.*) 6C
important (*adj*) 7A
impossible (*adj*) 6B
in (*prep*) 3A
increase (*v*) 7C
indefinite (*adj*) 5C
information (*n.m.*) 6A
injector (*n.c.*) 3B
instruction (*n.c.*) 4A
instrument (*n.c.*) 1C
into (*prep*) 5C
ion (*n.c.*) 6A
its (*adj*) 7A

jack (*n.c.*) 3B
jacket (*n.c.*) 3A
join (*v*) 4A

kettle (*n.c.*) 6C
kilowatt (kW) (*n.c.*) 6C
knife (*n.c.*) 1B
know (*v.*) 7A

label (*n.c.*) 1B
label (*v*) 2B
large (*adj*) 4C
last (*adj*) 2C
lathe (*n.c.*) 4A
least (*adj & adv*) 7B
leather (*n.m.*) 1B
left (*on the* left) 3B
length (*n.c. & n.m.*) 2C
less (*adj & adv*) 2B
lid (*n.c.*) 1B
life (*n.c.*) 6B
light (and *head*light) (*n.c.*) 3A
light (*n.m.*) 5A
light (*adj*) 8A
line (*n.c.*) 5B
lining (*n.c.*) 6A
liquefy (*v*) 5A
liquid (*n.c. & n.m.*) 4B
little (*a* little) (*adj & pron*) 3C
lock (*n.c.*) 1B
long (*adj*) 2C
loosen (*v*) 4B
lorry (*n.c.*) 3C
lot (*a* lot *of*) (*adj & pron*) 3C
low (*adj*) 5A
lower (*v*) 4B

machine (*n.c.*) 4A
made (of) 1B
magnet (*n.c.*) 5A
main (*adj*) 4C
mains (*n.m.*) 4B
maintenance (*n.m.*) 6C
make (*v.*) 4A
malleability (*n.m.*) 7A
malleable (*adj*) 7A
mallet (*n.c.*) 1A
manufacture (*v*) 6B
many (*adj & pron*) 8A
mark (*v*) 2B
material (*n.c.*) 1B
matter (*n.m.*) 5A
maximum (*adj*) 8C
measure (*v*) 1C
measurement (*n.c.*) 7B
mechanical (*adj*) 8B
medium (*adj*) 7B
melt (*v*) 5A
melting point (*n.c.*) 5A
metal (*n.c. & n.m.*) 1B
metallic (*adj*) 8A
micrometer (*n.c.*) 1C
microphone (*n.c.*) 1C
microscope (*n.c.*) 1C

middle (*in the* middle *of*) 3A
millilitre (ml) (*n.c.*) 5B
minimum (*adj*) 8C
minus (*adv*) 5A
mirror (*n.c.*) 3B
mix (*n.c.*) 8C
mixture (*n.c.*) 3C
molecule (*n.c.*) 5C
more (*adj & adv*) 2B
most (*adj & adv*) 3C
motion (*n.c. & n.m.*) 5C
motionless (*adj*) 5C
motorcycle (*n.c.*) 3A
move (*v*) 5C

nail (*n.c.*) 1C
name (*n.c.*) 1A
near (*prep*) 4B
negative (*adj*) 6A
never (*adv*) 2B
new (*adj*) 6B
next (*adj & adv*) 2C
non- 8A
nonagon (*n.c.*) 2B
none (*pron*) 3C
nor (*conj*) 7A
normal (*adj*) 5A
normally (*adv*) 5A
not at all (*adv*) 7A
note 4C
now (*adv*) 6B
nozzle (*n.c.*) 4C
number (*a* number *of*) 8C
nut (*n.c.*) 1C

object (*n.c.*) 6A
obtuse (*adj*) 2B
octagon (*n.c.*) 2B
octagonal (*adj*) 2C
often (*adv*) 4C
oil (*n.c. & n.m.*) 1B
on (*prep*) 3A
only (*adv*) 4C
open (*adj*) 5C
open (*v*) 4C
opposite (*adj*) 2A
or (*conj*) 1A
original (*adj*) 7A
other(s) (*adj & pron*) 3A
oval (*adj*) 2C
over (*prep*) 3B

paper (*n.m.*) 1B
parallel (*adj*) 2A
parallelogram (*n.c.*) 2A
part (*n.c.*) 1A
particle (*n.c.*) 8C
pentagon (*n.c.*) 2B
pentagonal (*adj*) 2C
per (*prep*) 5B
per cent (*adv*) 3C
percentage (*n.c.*) 3C
petrol (*n.c. & n.m.*) 1B
picture (*n.c.*) 1A
pin (*n.c.*) 4C
pipe (*n.c.*) 2C
place (*n.c.*) 4B
place (*v*) 4B
plane (*n.c.*) (i) 1A
(ii) 2A
plastic (*n.c. & n.m.*) 1B
plate (*n.c.*) 6A
plate (*v*) 6A
plating (*n.m.*) 6A
pliers (*n*) 4C
plug (*n.c.*) 4C
plus (*adv*) 6A
point (*n.c.*) 4B
point (*v*) 4A
polished (*adj*) 6C
polygon (*n.c.*) 2B
poor (*adj*) 7B
port (*n.c.*) 3B
position (*n.c.*) 5C
positive (*adj*) 6A
possible (*adj*) 6B
power (*n.m.*) 6C
press (*v*) 4C
pressure (*n.c. & n.m.*) 1C
primary (*adj*) 6B
prism (*n.c.*) 2C
process (*n.c.*) 6A
produce (*v*) 6A
production (*n.m.*) 8B
property (*n.c.*) 7A
proportion (*in* proportion *to*) 7C
protractor (*n.c.*) 1C
provide (*v*) 6A
pure (*adj*) 5B
purify (*v*) 6A

quickly (*adv*) 8B
quite (*adv*) 7A

radiate (*v*) 5A
radiator (*n.c.*) 3A
radius (*n.c.*) 4B
rain gauge (*n.c.*) 1C
raise (*v*) 4B
range (*n.c.*) 8B
rapidly (*adv*) 5C
rare (*adj*) 8A
ratio (*n.c.*) 8C
read (*v*) 4A
recharge (*v*) 6B
rectangle (*n.c.*) 2A
rectangular (*adj*) 2C
rectilinear (*adj*) 2A
red (*adj*) 4C
reflect (*v*) 5A
reflector (*n.c.*) 6C
reflex (*adj*) 2B
refract (*v*) 5A
refractoriness (*n.m.*) 7C
refractory (*adj*) 7C
regain (*v*) 7A
regular (*adj*) 2B
release (*v*) 6B
remain (*v*) 5C
remove (*v*) 4B
replace (*v*) 4B
require (*v*) 6C
resist (*v*) 6C
resistance (*n.c.*) 6C
resistive (*adj*) 7C
resistivity (*n.m.*) 7C
retain (*v*) 7A
rhombus (*n.c.*) 2A
right (*on the* right) 3B

water-cooled (*adj*) 3A
watt (W) (*n.c.*) 6C
wax (*n.m.*) 5B
way (*in the same* way) 8C
weak (*adj*) 7C
weight (*n.c. & n.m.*) 1C
well (*n.c.*) 2C
well (*adv*) 6C
what . . . ? (*adj & pron*) 1A
wheel (*n.c.*) 3A
where . . ? (*adv*) 4A
which . . ? (*adj & pron*) 7B
white (*adj*) 4C
why . . ? (*adv*) 6C
wide (*adj*) 2C
width (*n.c. & n.m.*) 2C
windscreen (*n.c.*) 3B
wiper (*n.c.*) 3B
wire (*n.c. & n.m.*) 6C
with (*prep*) 3C
withstand (*v*) 7C
without (*prep*) 4C
wood (*n.m.*) 1B
wooden (*adj*) 1B
work (*n.m.*) 8C
workbench (*n.c.*) 4A
workshop (*n.c.*) 4A
worse (*adj*) 8B
worst (*adj*) 8B
wrench (*n.c.*) 1A

zero (*adj & n.c.*) 5A